Nefedov, V. I. (Vadim Ivanovich)

X-ray photoelectron spectroscopy of solid surfaces

X-ray Photoelectron Spectroscopy of Solid Surfaces

X-ray Photoelectron Spectroscopy of Solid Surfaces

Vadim I. Nefedov

Institute of General Inorganic Chemistry,
Academy of Sciences of USSR, Moscow, USSR

Translated by N. G. Shartse

Utrecht, The Netherlands
1988

VSP BV (formerly VNU Science Press BV)
P.O. Box 346
3700 AH Zeist
The Netherlands

First English edition published 1988

CIP-DATA KONINKLIJKE BIBLIOTHEEK, DEN HAAG

X-ray photoelectron spectroscopy of solid surfaces/
V. I. Nefedov.—Utrecht: VSP.—Ill.
ISBN 90-6764-080-8 bound
SISO 533.5 UDC 539.26
Subject heading: solid surfaces; spectroscopy.

Phototypesetting by Thomson Press (India) Limited, New Delhi.

Printed in Great Britain at the Alden Press Ltd., Oxford

Contents

Preface

At present, X-ray photoelectron spectroscopy (XPS) also known as Electron Spectroscopy for Chemical Analysis (ESCA) is widely used to study solids surfaces and structure, though as recently as the 1960s, only a handful of laboratories employed it.

X-ray photoelectron spectroscopy can be applied to all the elements and used to make a quantitative analysis of solids surfaces, including monolayers. It is employed extensively to study oxidation, adsorption, catalysis, corrosion, diffusion, and also thin films and coatings. In addition, X-ray photoelectron spectroscopy helps to establish functional groups and the oxidation state of an element as well as to evaluate its effective charge in compounds under study.

The monograph has two objectives. One is to meet the interests of both practicing and budding specialists in the field. Therefore Chapters 1–3 attempt a systematic presentation of the physical principles of the method. This part is intended for readers altogether unfamiliar with the method. These chapters can therefore be used as a textbook by those who plan to thoroughly and professionally master X-ray photoelectron spectroscopy.

The other is to meet the needs of research and industrial personnel who would like a glimpse of the potentialities of that method and the results obtained with its help in various practical fields. Chapters 4 and 5 therefore describe a selection of the more important findings of X-ray photoelectron spectroscopy research into catalysts, corrosion, oxidation of solids, adhesion, electron surface states and other important fields, as can be gained from the contents.

The 1983 Russian-language edition has been thoroughly revised and supplemented with more than 180 new references updated till the end of 1985, while several new sections have been added.

As a companion volume to this work, a book covering an equally important method of surface analysis entitled *Secondary Ion Mass Spectroscopy of Solid Surfaces* by V. T. Cherepin is available from the same publisher.

V. I. Nefedov

Chapter 1

Binding energy of inner electrons and identification of chemical compounds

1.1. Physical principles and experimental techniques of X-ray photoelectron spectroscopy

The theoretical foundation of X-ray photoelectron spectroscopy (XPS) was laid way back in 1905, when Einstein wrote his well-known photoeffect equation:

$$h\nu = E_b + E_{kin} \tag{1.1}$$

where $h\nu$ is quantum energy, E_b the binding energy of the electron in matter and E_{kin} the kinetic energy of the ejected electron. The method of X-ray photoelectron spectroscopy [1–17] can be summed up as the measurement of the kinetic energy of the inner or valence electron ejected by an incident photon with a known energy $h\nu$. Knowing these values, it is easy to calculate E_b, the binding energy, which is a sensitive characteristic of chemical bonds in a compound and is related to the energy of ionization.

X-ray photoelectron spectroscopy analysis is applicable to all the elements except hydrogen in solids, liquids, and gases. Since the experiment is easier to perform on solids in the form of powder or plates, preference is usually given to low-volatility samples, while high-volatility substances are frozen. The minimum sample weight that can be analysed is 10^{-5} g, while a mere 10^{-7} to 10^{-9} g of an element can be detected. Usually, samples of 10–100 mg are taken. They can be cooled to the temperature of liquid nitrogen or heated to several hundred °C. References [12] describe experiments in which samples were cooled to the temperature of liquid nitrogen or heated to several thousand °C. The main constraints on samples are as follows: (1) the substance must be stable in vacuum*; and (2) the substance must not disintegrate substantially under the effect of X-ray radiation. Most substances meet these demands.

Usually, the escape depth of ejected electrons is not larger than 5 nm†; there must therefore be certainty that the 5 nm-thick surface layer adequately represents the bulk of the substance in question. At the same time, this circumstance is conducive to surface analysis. If the analysed surface layer and its base are chemically different, even monoatomic layers (including absorbed molecules) can be studied.

---

*Vacuum of at least 10^{-3} Pa, usually 10^{-4} to 10^{-9} Pa.

†See Chapter 2, Section 2.2 for details.

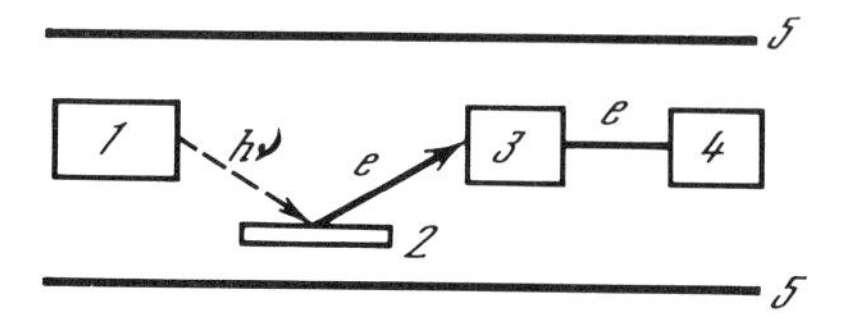

Figure 1.1. Skeleton diagram of the X-ray photoelectron spectrometer.
1 = X-ray rube; 2 = sample; 3 = photoelectron spectrometer; 4 = detector; 5 = magnetic shield.

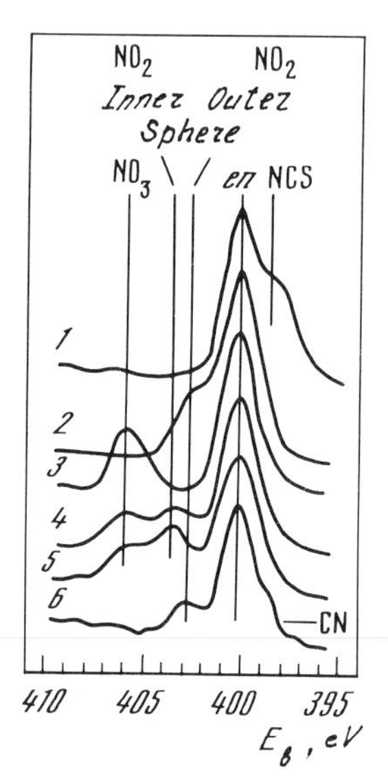

Figure 1.2. N 1s [4] lines.
1 = [Iren$_3$] (NSC)$_3$; 2 = [Iren$_3$] (NO$_2$)$_3$; 3 = [Iren$_3$] (NO$_3$)$_3$; 4 = [Iren$_2$CINO$_2$] (NO$_3$); 5 = [Iren$_2$(NO$_2$)$_2$] (NO$_3$); 6 = [Co(NH$_3$)$_5$CN] (NO$_2$)$_2$.

Figure (1.1) is the skeleton diagram of an experimental XPS unit. X-rays from the tube fall on the sample positioned next to the slit of the spectrometer and eject electrons from inner and valence levels. The ejected electrons pass through the slit into a high-resolution electron spectrometer, which measures their kinetic energy. A focused monochromatic electron beam is received by the electron detector. The spectra are represented in Fig. (1.2), which shows a number of their important characteristics: the area under the curve is proportional to the number of the ejected electrons, i.e. to the number of atoms of this type in the molecule, while the peak position is typical of every functional group. The spectra are denoted in the following manner: the chemical symbol of the element is followed by the notation of the electron level from which the electron was ejected, e.g. Pt $4f_{7/2}$, Nls, etc. The value of the quantum number j is sometimes omitted for simplicity's sake and Pt 4f, for instance, stands either for the spin-doublet Pt $4f_{5/2}$ and Pt $4f_{7/2}$ or for the more intensive Pt $4f_{7/2}$ line.

An idea of the basic elements of the apparatus and conditions under which spectra are obtained is necessary to understand the method and its potentialities.

X-ray photoelectron spectrometers are produced by the Soviet Union and by a number of Western companies, among them Vacuum Generators, Hewlett-Packard, McPherson, and AEI. Company publications contain sufficient information on these instruments.

1.1.1. *X-ray sources*

Conventional X-ray tubes serve as X-ray sources. The tubes are used at $V = 8-15$ kV and $J = 20-200$ mA, depending on the spectrometer type. Rotating-anode tubes also are employed to prevent local anode overheating. Anodes are usually made of Al or Mg, the energies of the K_α-lines of which equal 1486.6 and 1253.6 eV, respectively. The K_α-lines of F, Si, Cu, Cr, and the Mξ-lines of Y and Zr are also used.* The X-ray tube and the sample are divided by a thin film, through which the X-rays passes. The tube and the sample have to be divided to prevent scattered electrons from the tube from penetrating into the sample chamber.

If monochromatic X-ray radiation is used to eject electrons from the sample, the resolution of the instrument is improved (see below). Thin filters are sometimes placed between the tube and the sample for this purpose, or curved crystals are employed (e.g. in a Swedish instrument and in the Hewlett-Packard spectrometer). Monochromatic radiation also helps to simplify the X-ray electron spectrum. There are $\alpha_{3,4}$ satellites close to the $K_{\alpha_{1,2}}$-line, separated by about 10 eV for Mg and Al, which duplicate the entire spectrum from the $K_{\alpha_{1,2}}$-line. The intensity of these secondary lines is 8 and 4 per cent of the basic line of the X-ray electron spectrum.†

The following method [20] of exciting photoelectron spectra deserves special mention as it makes it possible in principle to scan specimens with spatial resolutions of around 20 μm. A thin layer of specimen material (0.1–1 μm) is deposited on one side of a thin Al foil (some 6 μm thick). A highly focused electron beam is directed at the other side of the foil so that the AlK_α-line is excited. Since the X-ray intensity speedily decreases with the increase of the angle between the normal to the film surface and the photon flux, photoelectrons are excited predominantly in a very small part of the specimen, with dimensions comparable to the beam diameter on the reverse side of the Al foil. Reference [21] describes the possibility of obtaining photoelectron spectra from spots with a diameter of 250 μm. Synchrotron radiation is another source of continuous (and polarized) radiation with energies of up to 350 eV or more. It is especially good in studying valence levels.

Electron guns can be used alongside X-rays to eject electrons. The advantage offered by excitation of this type is that the electron beam is easily focused and that excitation energy can be constantly varied. Electron excitation, however, has two substantial drawbacks: (1) the considerable scattering of falling electrons

*Radiation of the He (I) and (II) lines with energies of 21.2 and 40.8 eV is also employed to study valence levels in gases and solids; these spectra are referred to as photoelectron spectra. The subdivision of electron spectra into X-ray and photoelectron spectra depending on excitation energy is conventional, rooted in the evolution of the method.

†See references [18, 19] for a detailed discussion of the energy status and relative intensity of the K satellites of the Al K_α- and Mg K_α-lines.

tangibly increases background radiation, especially in the case of solids; (2) most of the substances of interest to chemical research are decomposed by electron bombardment.

1.1.2. *Shielding from the geomagnetic field*

Since the progress of electron beams in the spectrometer (monochromator) depends on magnetic fields and since it is necessary to ensure that electron trajectories depend only on the field of the (magnetic or electrostatic) monochromator, the geomagnetic field near the monochromator must be reduced virtually to zero. The techniques of compensating for the geomagnetic field depend on electron energies and the monochromator type but usually they can be reduced to one of the following methods or their combination: (1) a set of Helmholtz coils; (2) a mu-metal paramagnetic screen. The latter technique prevails today. Mu-metal cannot be used in magnetic monochromators because a paramagnetic screen not only cuts off the geomagnetic field but also disturbs the field of the spectrometer, thus detracting from its resolution.

1.1.3. *Vacuum system*

The vacuum in the spectrometer chamber, depending on the type of instrument, between is 10^{-4} and 10^{-9} Pa (usually 10^{-6}–10^{-8} Pa). When the adsorption of pure metal surfaces or electron surface states is studied, a vacuum of some 10^{-8} Pa is necessary because otherwise the surface will be contaminated with residual gases. Estimates show [12] that with a 10^{-7} Pa vacuum, O_2 molecules with an adhesion coefficient equalling unit take 50 min to cover the surface with a monomolecular layer. Since time stands in reverse proportion to pressure, at 10^{-6} Pa vacuum the monomolecular layer takes a mere 5 min to form, which is comparable to spectrum recording time.

When conventional specimen surfaces—i.e., surfaces formed by some reactions under conventional conditions—are analysed, vacuum requirements are far less stringent because the adsorption of residual gases in the spectrometer does not substantially alter the composition of the surface film of a specimen exposed to the atmosphere. (Only a layer of hydrocarbons can be adsorbed additionally.) Hence the surfaces of real objects can be studied in a vacuum of only 10^{-4} Pa. However, if ion sputtering is used (see Chapter 3), the vacuum must be sufficiently high in most cases as well. A high vacuum in the spectrometer is obtained by diffusion pumps in the AEI and VG instruments, by ion pumps in the instruments of the SKB AP (USSR Academy of Sciences), Du Pont, Varian, and Hewlett-Packard, and by turbomolecular pumps in instruments produced by McPherson and in some of Varian models. Forevacuum is usually created by mechanical pumps. Adsorption pumps may help reduce the contamination of the specimen with vapours of the pumping fluid.

1.1.4. *Monochromators*

Monochromators of various types are used in X-ray photoelectron spectroscopy for electron focusing [15, 16].

1.1.5. *Magnetic monochromators*

Monochromators of this type use a magnetic field for electron focusing. The double focusing theory and the first instruments using this principle were developed by Prof. K. Siegbahn's team [2, 3]. These instruments are iron-free, as they are usually made of aluminium or bronze. Spectrometers with a radius of 300 mm are sufficient for most chemical studies. Double focusing is ensured by an inhomogeneous magnetic field, created by a set of four cylindric coils. Siegbahn's team also used a permanent magnet to focus electrons [2].

1.1.6. *Electrostatic monochromators*

Monochromators of this type are widely used in serial spectrometers. Figure (1.3) illustrates the design of the serial Varian monochromator. It incorporates a spherical condenser, which focuses electrons with definite energies at the exit slit. Electrons do not enter the monochromator immediately from the source: there is a retarding field, E_{ret}, between the spectrometer and the source, which reduces electron energies from roughly 1000 to 10–100 eV. This makes it possible to use a smaller monochromator and to increase signal intensity. Spectrum energy scanning is done by a computer-controlled retarding field. The kinetic energy of electrons entering the monochromator (analyser), E_{an}, does not change in the course of measurement. Since the values E_b and E_{an} are bound by the equation

$$E_b = h\nu - E_{an} - E_{ret}, \qquad (1.2)$$

when E_{ret} is defined, the value of E_b for the electrons penetrating into the detector is fully determined.

For some spherical analysers of this type, for which the energy of electrons

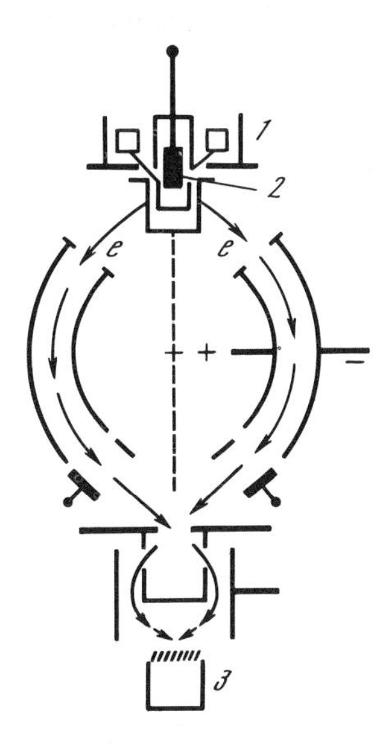

Figure 1.3. Schematic illustration of the VIEE-15 Varian spectrometer. 1 = X-ray tube; 2 = specimen; 3 = photoelectron multiplier.

entering the analyser is constant, the instrumental broadening of spectrum lines width remains constant as well. These analysers are used in the instruments developed by the SKB AP of the USSR Academy of Sciences, and also in Vacuum Generators and Hewlett-Packard analysers.

The spectrum can be scanned in some sectoral spherical analysers in a different way: the field on the condenser grids can be varied successively to feed electrons of different energies to the entrance slit. The instrumental line-width will be altered in this case: the higher the electron energy E_{an} in the analyser, the broader the peak because the value of the instrumental line-width stands in rough proportion to the value of $E_{an}^{3/2}$. These analyzers are used in McPherson spectrometers. AEI instruments incorporate both methods of spectrum scanning.

PHI instruments uses a cylindrical mirror analyser. This has two coaxial cylinders, the internal one being negatively charged. The specimen and detector are positioned on the axis of the cylinders. Electrons enter the analyser after passing a retarding field. The spectrum is scanned by varying the potential on the external cylinder or by altering the retarding field. All the analysers described above are known as deflection or dispersion analysers.

The Du Pont instrument uses an analyser that combines dispersion and non-dispersion principles. First, a band of electron energies is singled out in a low-resolution dispersion-type analyser. These electrons pass through a low-velocity filter, which deflects only electrons with energies below a certain level. These deflected electrons are focused and directed to a high-velocity electron filter, which allows only electrons with energies higher than the filter nominal energy to reach the detector. Electrons in a narrow energy band are singled out in this way.

Let us now consider a question of major importance to the analysis of specimen composition on the basis of XPS spectrum intensity, namely, the dependence of the efficiency T of different analysers on the kinetic energy E_{kin} of different photoelectrons. It should be noted first and foremost that analysers with retarding fields make it possible to achieve, with comparable resolution values, a higher spectrum intensity than analysers without retarding fields [12, 22]. Indeed,

$$T \sim BA\Omega, \tag{1.3}$$

where B is the luminosity of the electron source per unit of area and unit of the body angle; A is the area of the electron source; and Ω is the body angle electron exit from the source to the analyser (Fig. 1.4).

If the analyser resolution is denoted ΔE_{kin}, then, in the case of the retarding field,

$$B_0 = B(E_{an}/E_{kin}), \quad A_0 \sim (\Delta E_{kin}/E_{an})^{3/2},$$
$$\Omega_0 \sim (\Delta E_{kin}/E_{an})^{1/2}, \tag{1.4}$$

and without the retarding field

$$A \sim (\Delta E_{kin}/E_{kin})^{3/2}, \quad \Omega \sim (\Delta E_{kin}/E_{kin})^{1/2}. \tag{1.5}$$

Hence,

$$T_0/T = E_{kin}/E_{an}, \tag{1.6}$$

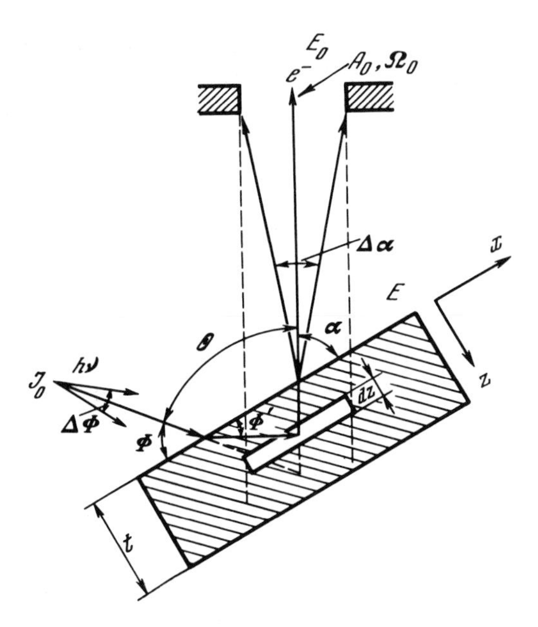

Figure 1.4. Schematic illustration of an XPS spectrometer.

i.e. $T_0 > T$. Since the values A_0 and Ω_0 in the spectrometers with the retarding field are fixed by the instrument design, in these analysers

$$T_0 \sim E_{an}/E_{kin}, \tag{1.7}$$

which means that intensity diminishes with a decrease in E_{an} and grows with smaller values of E_{kin}. As E_{an} diminishes in accordance with (1.4), ΔE_{kin} decreases, that is, resolution improves, but the signal/background relationship decreases [see Table (1.1) for the Varian analyser].

The theoretical equation (1.7) often does not hold and the power denominator n with E_{kin} is often about $n = -\frac{1}{2}$. This holds true, for instance in the Hewlett-Packard and Vacuum Generators spectrometers [23].

In the spectrometers in which electron energy in the analyser is not constant,

$$T \sim E_{kin}, \tag{1.8}$$

i.e. line intensity grows as the E_{kin} value increases.

1.1.7. *Detectors*

Usually channel electron multipliers are used as electron-counting detectors. The more advanced registration method is offered by the multi-channel miniature multiplier made by Hewlett-Packard. They count electrons in many channels according to different values of E_b. Since the intensity of the spectrum spot is usually measured in about 100 points, the application of the multi-channel multiplier, with the other conditions being equal, reduces spectrum registration time to about one-hundredth of the conventional value.

1.1.8. *Resolution*

This is one of the major characteristics of any instrument. Resolution depends on the value of the line width E_l in the XPS spectrum. The value E_l, measured in half the line intensity, equals

$$E_l = E_{lev} + E_{in} + E_{l.x} + E_{u.c}, \qquad (1.9)$$

where E_{lev} is the width of the analysed level; E_{in} is instrument broadening caused by the impossibility of the ideal focusing of electrons; $E_{l.x}$ is the width of the X-ray line by which electrons are ejected from the specimen; and $E_{u.c}$ the broadening caused by the unequal charging* of powder particles (for non-conductive powders). Equation (1.9) is conventional. The half-width of the resulting line is the total of the half-widths of different components if the Lorenz distributions hold for the latter.

The value E_{lev} varies in a broad range from tens to tenths of eV. Since the width of inner electron levels grows as the transition is effected from peripheral to deeper levels, analysis is usually directed at inner levels, with the main quantum number one or two units below that of valence electrons—for instance, 1s-electrons for the first period, 2p-electrons for the second and third periods, 3d-electrons for elements of the Pd group, etc. In this case, usually $E_{lev} \simeq 1$ eV.

The value E_{in} in different instruments varies from 0.3 to 2 eV depending on spectrum scanning conditions, and amounts to about one per cent of E_{an} (Table 1.1).

The value $E_{l.x}$ equals 0.6–0.8 eV for the Mg and Al K_α-lines. The values $E_{l.x}$ is double or treble that for the Cr and Cu K lines. The value $E_{l.x}$ is roughly 40 per cent of E_{lev} even with the Al and Mg K_α-lines. That is why a combined focusing method (Fig. 1.5) 2 has been developed, making it possible with the help of a crystalline monochromator of X-rays and a magnetic (or electrostatic) field to exclude the contribution of the natural width $E_{l.x}$ of the X-ray excitation line to the value E_{lev}. The specimen in this instance is positioned at such an angle to the falling monochromatic X-ray beam that the photoelectron emitted by different areas of the specimen and having different kinetic energies are focused in the same point since the path for the electrons excited by the high-energy edge of the X-ray line and having the energy $E + E_{l.x}/2$ is longer than that for the electrons excited by the other edge of the line with the energy $E - E_{l.x}/2$ (Fig. 1.5). This method is efficient only for specimens with sufficiently smooth surfaces.

Table 1.1
Dependence of intensity I_{max} in line maximum, background intensity I_{back} and width E_l of the C ls line on E_{an}

E_{an}, eV	E_l, eV	I_{max}, imp s^{-1}	I_{back}, imp s^{-1}	I_{max}/I_{back}	$(I_{max} - I_{back})/I_{back}$
100	1.85	7500	1250	6	5
50	1.30	3100	600	5	4
20	1.05	1150	950	1.2	0.2

*Specimen charging effects are described in the next section.

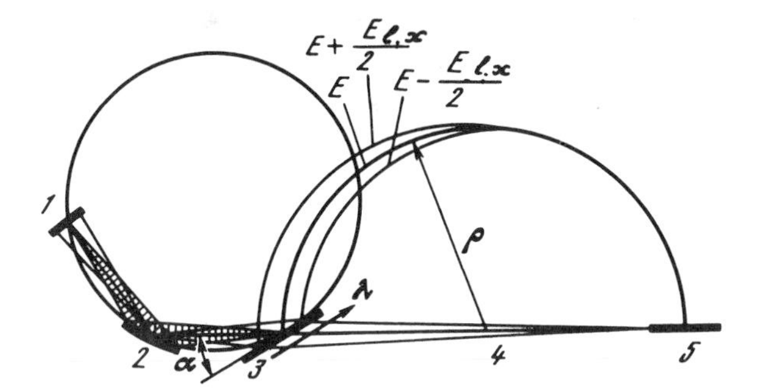

Figure 1.5. Combined focusing principle.
1 = X-radiation source; 2 = crystalline monochromator of X-radiation; 3 = specimen; 4 = electronic spectrometer; 5 = detector.

The value $E_{u.c}$ may reach several tenths of eV. To reduce it, we should obtain as smooth and uniform films of the analysed powder as possible on the surface of the specimen, or expose the specimen to the flow of slow electrons.

The value E_l therefore averages 2–3 eV at $E_{an} = 100$ eV in instruments without combined focusing. The results of the tests of spectrometers of different types show that the use of X-ray line monochromators with powder specimens reduces the value E_l by roughly 0.3–0.5 eV [24]. The resolution of the spectrometer today is usually enough to differentiate between atoms of the same element with different oxidation states in one compound (see Fig. 1.2).

1.2. Specimen charging and spectrum calibration

If the analysed specimen is electrically conductive, electrical equilibrium is achieved between the specimen and the material of the spectrometer (the Fermi levels are equalized). It is natural in the case of such specimens to assume that the Fermi level of the spectrometer material is equal to zero in determining the kinetic energy of the ejected electron, i.e. to compute the value E_b from the equation

$$E_b = h\nu - E_{kin} - \Phi_{sp}, \tag{1.10}$$

where Φ_{sp} is the work function of spectrometer material.

The relationship between different energy characteristics is presented in Fig. (1.6) [12], where Φ_0 is the work function of the electrically conductive specimen. When two conductors (metals) come into contact, electrons pass from the metal with a lower value to the other metal until the Fermi levels of both become equal. A potential difference $\Phi_0 - \Phi_{sp}$ is reached between the specimen metal and the spectrometer, increasing or decreasing the measured E_{kin} of the specimen electrons. In other words, vacuum levels of the specimen and the spectrometer are different, and it becomes necessary to add Φ_0 to E_b to obtain the ionization energies E_{ion} relative to the vacuum level:

$$E_{ion} = E_b + \Phi_0. \tag{1.11}$$

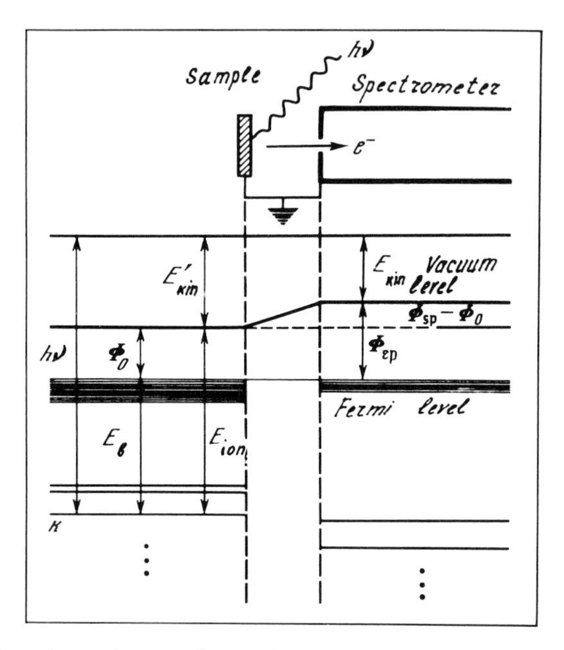

Figure 1.6. Interdependence of energy characteristics of conductive specimen and spectrometer.

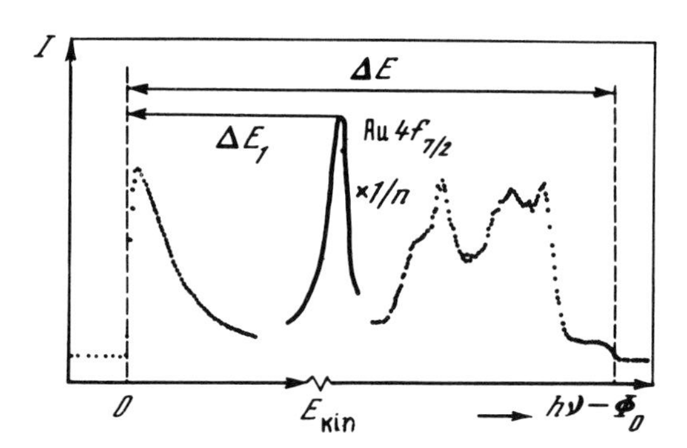

Figure 1.7. Appropos determination of absolute values of E_{ion}.

The spectrometer measures E_{kin} and it is sufficient to know the value $\Phi_0 - \Phi_{sp}$ to obtain the value E_{ion}:

$$E_{ion} = h\nu - E_{kin} + (\Phi_0 - \Phi_{sp}). \tag{1.12}$$

The value Φ_{sp} is usually considered known and the problem is reduced to the determination of Φ_0. Generally speaking, the values Φ_0 or E_{ion} can be determined directly from XPS spectra [25–27]. The principle of this operation is as follows: If the entire spectrum of metal photoelectrons is measured [see for example the Au spectrum in Fig. (1.7)], the Fermi level can be determined from the position of the line with the maximum kinetic energy. Every spectrum is accompanied by a band of secondary electrons obtained from a series of interactions of primary photoelectrons both with the specimen and with spectrometer material. The left-hand end of that band (Fig. 1.7) corresponds to zero kinetic energy of secondary electrons. If $\Phi_0 < \Phi_{sp}$, the spectrometer is charged negatively as regards the specimen and secondary electrons obtained from interaction with spectrometer material have the zero kinetic energy of the line edge. The kinetic energy of specimen secondary electrons, all other conditions being equal, is less by $e(\Phi_{sp} - \Phi_0)$ than that of spectrometer secondary electrons, so the latter can still reach the analyser, while the secondary electrons of specimen material will have zero energy.

In this way, with $\Phi_0 < \Phi_{sp}$, the left-hand end of the band of secondary electrons corresponds to zero kinetic energy of the spectrometer electrons, and with $\Phi_0 > \Phi_{sp}$ to zero kinetic energy of the specimen electrons. The latter result can always be obtained if the specimen is charged negatively with regard to the spectrometer, i.e. if the potential difference V is applied to the specimen and the spectrometer. In this case the following equations hold for the work function and E_{ion} of the Au $4f_{7/2}$ level (Fig. 1.7):

$$\Phi_0 = h\nu - \Delta E, \quad E_{ion} = h\nu - \Delta E_1. \tag{1.13}$$

Reference [27] states that the maximum values of ΔE_1 should be striven for in determining the E_{ion} and work function. Maxima are achieved with oblique angles of the fall of radiation on the specimen.

If the specimen is a semiconductor (or insulator), the Fermi level of the spectrometer is found between the last filled level and the conduction zone (or specimen vacuum level), which makes it substantially more difficult to calculate E_{ion} (see below). An additional difficulty is entailed by the possible difference between the electrical potentials of the specimen and the spectrometer material, although in principle we can expect an evening out of the potentials of the analysed insulator surface layer deposited on a conductive substrate and the spectrometer material if the insulator layer is sufficiently thin [2]. This situation, however, is only a borderline case which is rarely obtained in practice. It is known that the positive electrical charge accumulated on the irradiated specimen due to ejected electrons can shift atom levels by a few eV. The equation for E_b can be written in the general case:

$$E_b = h\nu - E_{kin} - \Phi_{sp} \pm E_{dif} \tag{1.14}$$

where E_{dif} is the difference between the potentials of the surface layer of the specimen and the spectrometer material. In the general case, the surface layer of the specimen can be charged either positively or negatively with regard to the spectrometer. A negative charge, in particular, can be induced by large numbers of secondary electrons close to the specimen surface.

References [28–33] and their bibliographies analyse different factors influencing the value E_{dif}, such as the voltage and current in the X-ray tube, the angle between the specimen and the X-ray flux, the secondary electron current, and the current between the specimen and the holder. These references also suggest some methods for measuring or evaluating E_{dif}. However, at present methods employing various standards are used to evaluate E_{dif}.

1.2.1. *Inner standard*

If we have in the analysed series of compounds an atom whose value E_b ought not to change substantially in that series, it is sufficient to rate all the measured values E_b by the same value E_b for that atom to determine E_b alterations in the analysed series. For instance, in analysing a series of $(PPh_3)_2PtX_2$ compounds, where X stands for different acidoligands, the value C 1s in the phenyl groups is taken for the standard value, equalling 285.0 eV with zero charge. If the measured value C 1s, according to (1.10), is equal to 284.0 eV, just 1.0 eV is added to the measured values Pt 4f to obtain their correct values. The binding energy of the substrate atom can also be used as the inner standard in studying various surface compounds or adsorbed molecules if there is certainty that the substrate is not modified substantially in the specimen.

1.2.2. *External surface standard*

Use has to be made of external standards in the general case. The most widespread external standard nowadays is the value C 1s (285.0 eV) of the hydrocarbon layer deposited on the specimen as a result of the infiltration of the spectrometer volume by vapours of diffusion oil, the degassing of organic seals, etc. Reference [34] analyses the mechanism of the formation of the hydrocarbon layer in greater detail. There are facts indicating that the charging of the hydrocarbon layer and the specimen is identical, which means that there is electrical equilibrium between the hydrocarbon layer and the specimen. First, the measured values E_b (minus E_{dif}), according to (1.10), can vary within ± 1 eV depending on the conditions of the experiment, but if we take account of the measured C 1s, the results are well reproduced (see for example [35, 36]). The results are well reproduced by instruments of different types as well [24, 35, 37]. Table (1.2) [35] exemplifies this point by a series of data for Na_2SiF_6 and ZnO compounds.

Second, the results obtained with the help of that standard do not depend on the difference between the potentials of the specimen and the spectrometer, neither do they depend on whether substances are deposited on an insulation tape or rubbed into the conductive sample-holder, though when a substance is deposited on an organic non-conductive film, the charge of the specimen exceeds the measured binding energy by 1 eV.

Table 1.2
Binding energies obtained with instruments of different types and reduced to C 1s = 285 eV

Instrument type	Substance	Na_2SiF_6			ZnO	
VIEE-15	MgKa + AlKa	1072.3	104.5	686.8	1021.9	530.8
VIEE-15	AlKa	1072.0	104.7	686.3	1021.6	530.8
H-P	AlKa	1072.4	104.0	686.6	1021.9	530.3
ES-100	AlKa	1072.5	104.5	686.5	1021.5	530.5
ESCA-3	AlKa	1072.3	104.5	686.9	1021.5	530.6

There are some innate shortcomings in the external standard methods. The presence of organic additions or carbon atoms in the substance itself distorts the C 1s line. It is also necessary to make sure that more than one monolayer of hydrocarbons is deposited on the specimen [38]. It cannot be ruled out either that the hydrocarbon layer can be polarized [39]. The latter problem, however, has been poorly studied so far. In addition to polarization—the alteration of the C 1s energy of the hydrocarbon layer with substrate—we should expect a small difference between the potentials of the hydrocarbon layer and the analysed specimen. In particular, reference [37] considered six substances in nine different spectrometers. Four measurements out of 136 recorded that the use of the C 1s line as the external-standard does not guarantee that full account is taken of the charging effect of the analysed specimen. Reference [40] reviews studies concerned with the use of the C 1s line as the external standard.

Deposited metal films can also be used as surface external standards [41]. Reference [41] shows that the use of the Au $4f_{7/2}$ and Pd $3d_{5/2}$ lines from thin Au and Pd films deposited on $BaSO_4$ in vacuum gave virtually identical results for the Ba $4f_{7/2}$ line after correction. It is noteworthy that the authors observed two peaks in the Au $4f_{7/2}$ line, one of them corresponding to Au in contact with the spectrometer and the other to Au in contact with $BaSO_4$. When a potential difference of $\pm$ 1.6 eV was applied, the peak corresponding to Au in contact with the spectrometer was shifted by precisely that value. The shift of the other peak was different.

Even the above finding shows that some caution is needed in using this standard. Reference [42] shows that the quantity of deposited metal influences the contact of the metal and the analysed specimen. If too much metal has been deposited, there can be no electrical contact between the insulator and the deposited metal and if too little metal is present, the deposited metal forms islets on the specimen which are not in electrical contact with the material of the spectrometer. Reference [43] recommends that an Au layer 0.6 nm thick be deposited, with a photoelectron exit angle of 45°. References [44, 45] show, furthermore, that a reaction between the deposited metal and the analysed specimen can be expected in some cases.

References [35, 46] draw a comparison between the use of the C 1s lines from a hydrocarbon layer and the Au 4f line from a deposited gold layer for a broad range of different chemical compounds. The findings obtained in [35] are given in Table 1.3. In most cases both spectrometers in this study yield similar results with the use of the C 1s line before and after Au deposition. After Au deposition, the values of the binding energies obtained with standardization using the C 1s and Au 4f lines coincide in a number of cases (NaF, WO_3), but these values are different from those before Au deposition, using the C 1s line as the standard. The explanation seems to be that the C 1s line is observed from the hydrocarbon layer on the surface of the Au layer, which is not in contact with the specimen under study.

References [47, 48] suggest that implanted Ar atoms should be used as the external standard. This method of correction is not applicable to most chemical compounds because they disintegrate under ion bombardment, but it is

Table 1.3
Binding energies of different spectrometers with different standards (eV)

Compound	VIEE-15	ESCA-3	ESCA-3	
	(C 1s = 285.0)	(C 1s = 285.0)	C 1s = 285.0	Au 4f = 83.8
	Before Au deposition		After Au deposition	
KCl	293.2; 198.5	293.0; 198.6	293.4; 198.9	293.4; 198.9
Al_2O_3	74.5; 531.4	74.1; 530.9	74.3; 531.0	74.1; 530.8
SiO_2	103.5; 532.7	103.5; 532.8	103.5; 532.9	103.2; 532.6
V_2O_5	517.6; 530.6	517.5; 530.7	517.7; 530.6	517.3; 530.2
WO_3	36.3; 531.0	36.0; 530.7	35.7; 530.2	35.6; 530.1
Fe_2O_3	710.7; 530.0	710.5; 529.7	710.5; 529.8	710.3; 529.6
$NiSO_4$	857.0; 532.4; 169.3	857.1; 532.1; 169.4	857.2; 532.4; 169.9	857.0; 532.2; 169.7
NaF	1071.4; 684.6	1071.6; 684.8	1072.4; 685.6	1072.8; 686.0
CdF_2	405.9; 684.4	405.6; 684.5	405.8; 684.7	405.5; 684.4
CuF_2	347.9; 685.0	348.1; 685.1	348.5; 685.3	347.9; 684.7
$NiF_2 \cdot 4H_2O$	857.7; 684.9; 533.2	857.7; 684.9; 532.8	857.9; 685.1 533.1	857.8; 685.0; 533.0
KNO_3	293.2; 407.4; 533.0	293.0; 407.5; 532.8	293.1; 407.7; 533.1	293.1; 407.7; 533.1
NaCl	1071.8; 198.4	1072.3; 199.1	—	
Na_2MoO_4	1071.9; 232.6; 530.7	1072.1; 232.4; 530.0	—	

undoubtedly useful in a number of special cases because the intensity of the C 1s lines after ion bombardment is virtually zero and they cannot be used to evaluate the specimen charge.

1.2.3. *External standard through admixture*

Reference [49] suggests that the F 1s line of LiF, added to the specimen substance, could be used as an external standard. References [39, 50, 51] also describe some attempts to use the lines of solids added to the specimen under analysis. Apparently, sound contact between insulators, or analysed insulators, and the conductor (if graphite is added), cannot always be secured even with thorough grinding [52]. In particular, the spectra of different mixtures of insulators [53] show that the position of XPS lines of substances in the mixture is the same as for a pure substance. For instance, two Al 2p lines and two C 1s lines from the hydrocarbon layer are observed in the mixture of $Al(OH)_3$ and Na_3AlF_6 in the same energy position in which they were observed in the pure substances. Hence the conclusion that the use of insulators as standards to account for the charging effect seems dubious, though it cannot be ruled out completely if specimens are prepared specifically and substances chosen efficiently.

Reference [54] suggests a method of achieving electrical equilibrium between the analysed non-conductive surface and the spectrometer with the help of a special electron emitter (using a current of 10^{-8} to 10^{-7} A). This method has become more or less widespread, though both the detailed study of this problem in reference [32] and verifications of the results obtained with its help [24] have proven it to be unreliable. It is noted in particular [24], that in different laboratories this method yields a considerable scattering of binding energy values for identical compounds and that, moreover, the values obtained markedly differ from those measured using other methods to compensate for the specimen charge.

Specimen charge compensation is the focal methodological problem in XPS. This problem, obviously, has been far from completely resolved. When insulators are analysed, the use of the C ls line from the hydrocarbon layer can be recommended at present. If this line is absent, a measurement with regard to the Fermi level should be introduced in analysing a conductor. The measurement of the Fermi level of nickel [55] can be recommended, since its energy position can easily be determined experimentally.

Another question closely related to the problem of specimen charging is that of the relationship between the Fermi level of the spectrometer material and its position in the energy gap of the semiconductor under electrical equilibrium, and also the factors influencing that position. These questions have been considered in many studies (see [56–60] and references cited in them). Let us consider the results obtained in [58], which studied the Si 2p line in n- and p-type silicon. The concentrations of P and B atoms were 2×10^{-19} and $4 \times 10^{-19}\,cm^{-3}$, respectively. Since the position of the Fermi level in the energy gap depends on the character and concentrations of the doping atoms, we should expect alterations in the position of the Si 2p-lines for silicon of the n- and p-types with regard to the spectrometer Fermi level. The Fermi level for p-type silicon with the given concentration of B atoms is at the top of the valence zone, and for n-type silicon at the bottom of the conductivity zone. The width of the energy gap in Si equals 1.1 eV. It was found that the value of Si 2p in n-type silicon is 1 eV higher than it is in p-type silicon. We should stress that the difference between binding energies only reflects the difference between the charges of silicon specimens, i.e. variations of the contact difference of potentials between the spectrometer and n- or p-silicon. If we add the value to E_b [see Equation (1.11)], we will obtain virtually identical values of E_{ion} for Si 2p, which reflects the practical identity of Si atoms in n- and p-silicon. At present the shift of binding energy is used extensively to determine the positions of the Fermi level (see for example [60, 61]). Eleven semiconductors of the A^2B^6 and A^3B^5 types were studied in reference [59]. It was found that the Fermi level of a thin deposited Au film lies in the middle of the energy gap in six semiconductors, at the top in two and at the bottom of the energy gap in three. Reference [56] illustrates for six A^3B^5 semiconductors that the Fermi level of Au lines moves 0.4 eV towards larger binding energies from the valence surface states in the energy gap.

If there are hydrocarbon admixtures, the Fermi level of the spectrometer lies approximately in the middle between the last occupied hydrocarbon level (about 13 eV) and the vacuum level because the binding energy of C 1s equals about 285 eV and the ionization energy of C 1s in gaseous hydrocarbons is about 290–291 eV. Presuming that the vacuum levels of the hydrocarbon layer and the specimen are close, it is sufficient to add 5–6 eV to the E value to obtain the E_{ion} value of the specimen. The value obtained in this way does not seem to differ from the real E_{ion} by more than 1–2 eV. The relative accuracy in substituting E_b for E_{ion}

in different compounds seem to be far higher with the use of the C 1s line as the standard. Indeed, the analysed chemical compounds are usually covered with a hydrocarbon film of several monolayers. (This film survives for the most part even in high vacuum unless special surface cleaning techniques are employed.) If we assume that these films do not influence the work function on the surface of the analysed specimen, the difference $\Delta E_{ion} - \Delta E_b$ where $\Delta E_{ion} = E_{ion}(1) - E_{ion}(2)$ and $\Delta E_b = E_b(1) - E_b(2)$, equals $\Phi_0(1) - \Phi_0(2)$. Considering that the work function is fully determined by the film, $\Delta E_b = \Delta E_{ion}$. The latter equation seems to hold with an accuracy of the order of 0.5 eV.

Most of the studies concerned with insulators cite E_b data measured with regard to the Fermi level of the spectrometer with C 1s equalling 285 eV. As distinct from the case with conductor specimens, this zero is conventional because there is no electron current between the spectrometer and the specimen surface (see [62] for more detail), or to be more precise, because in the measurement of electron kinetic energies in the spectrometer the difference between the potentials of the specimen and the spectrometer is fully (or almost fully) determined by the charge of the specimen due to various factors not related to the work function of the spectrometer. Finally, the rather conventional character of the value C 1s (285 eV) should be stressed. According to (1.10) and (1.12), the value C 1s of the hydrocarbon layer on different metals must depend on the work function of those metals. Indeed, the measured values of C 1s [63] equal 284.69 eV for Au, 284.66 eV for Pd, 284.68 eV for Ag, 284.92 eV for Cu, and 285.58 eV for oxidized Mg (E_b relative to the Pd Fermi level). The use of any constant value of C 1s for spectra calibration thus makes it possible to measure only the $\Delta E_{ion} \approx \Delta E_b$ values, where E_b is determined with the help of the C 1s line.

Present-day literature cites vastly different values of E_b for the same compounds. Difficulties involved in taking account of specimen charging created the impression that this inconsistency of values was caused primarily by that effect. The finding of four Round-Robins [24, 35, 37, 55] using different instruments and methods of specimen charge correction shows that the scattering of values is often explained by incorrect spectrometer calibration. Article [55]

Table 1.4
Binding energies (eV)

Line	E_{med}	Range	Root-mean-square deviation
Au $4f_{7/2}$	84.0	83.2–85.4[a]	0.35[a]
Au $4d_{5/2}$	335.2	334.1–336.5[b]	0.44[b]
Au $4p_{3/2}$	546.4	545.1–547.8	0.43
Cu $2p_{3/2}$	932.6	931.3–933.9	0.51
Cu $3p_{3/2}$	75.1	74.2–75.7	0.30
C1s(on Au and Cu)	284.7	283.8–285.6[c]	0.40[c]
CuL_3VV[d]	918.7	917.8–919.7	0.42

[a]Calculated without counting one 82.0 eV measurement.
[b]Ditto, 340.3 eV.
[c]Ditto, without two measurements, 286.7 and 283.0 eV.
[d]Kinetic energy.

Table 1.5
Binding energies (eV) according to the literature

Line	[64]	[65]	[52]	[63]	[66]
Au $4f_{7/2}$	84.07	84.0	83.8	83.98	83.98
Pd $3d_{5/2}$	335.20	335.2	335.2		
Ag $3d_{5/2}$	368.24		368.2	368.21	368.27
Fe $2p_{3/2}$	706.82				
Pt $4f_{7/2}$			71.0		
Cu $2p_{3/2}$	932.53	932.7	932.8	932.66	932.67
Cu 3p		75.25			75.14
Cu 3s		122.55	122.9		
$CuL_3M_{4.5}M_{4.5}$	918.65[a]	918.35[a]	918.3[a]	567.97[b]	567.97[b]

[a]Kinetic energy.
[b]$R\upsilon - E_{kin}$.

considers the data obtained with 38 spectrometers of 8 different makes. The measurement of the binding energies of pure metals after ion cleaning was studied. The conditions of measurement precluded strong charging effects and coincidence of data within ± 0.3 eV should have been expected with correct spectrometer calibration. This was not the case. The median energy values obtained and the range of measured values are given in Table (1.4). The average values E_{med} closely correspond [55] to the findings of special measurements of E_b after thorough spectrometer calibration (Table 1.5) and can be used to verify spectrometer calibration. Binding energies of compounds cited in Tables 1.2 and 1.3 can also be used for the purpose. The values E_b cited in the first column of Table (1.3) were obtained with the spectrometer which yielded in the Round-Robin [55] E_b values actually coincident with E_{med} values.

The charging effect is accounted for in different ways in the original papers, which in itself can result in measurement differences of up to 2 eV. Reference [14] analyses the data cited in several hundred papers and suggests corrections which make it possible to convert roughly 9500 E_b values in 4500 compounds to the standard C 1s = 285 eV.

1.3. Patterns in E_b values and identification of chemical compounds

The principal experimental value determined by XPS is the binding energy* of electrons E_b, which by definition is the energy needed to eject the electron from a certain electron level of the analysed substance to the Fermi level of the spectrometer material [see Equation (1.10)]. Special interest is shown in most cases not in the absolute value E_b but in its change ΔE_b for the electron level of the same element in different compounds. This value—ΔE_b, the energy level shift—is usually measured with regard to the free element (chemical shift). The value E_b is only used for solids; for gases the value E_{ion} is always cited. The average accuracy

*There are several inner electron levels in the atom, all of them shifting in the same direction by a roughly similar value (differences measure up to 10 per cent) in a compound. It is therefore sufficient to study any single level.

of the experimental values $E_{b(ion)}$ is $\pm(0.1-0.2)$ eV for solids and about ± 0.04 eV for gases.* The values E_b can be used to determine qualitative composition of the sample because they are substantially different for different elements. The rough values E_b for different electron levels in elements are cited in monographs [2, 11]. More accurate E_b values for elements are cited in the supplement (also see references [10, 14]).

Let us consider basic experimental regularities in E_b values to determine the character of chemical compounds. It will be illustrated below that the E_b shift changes roughly linearly along with the effective charge of the analysed atom. This makes it easy to explain most of the regularities described below.

1.3.1. *Dependence on oxidation state*

The early studies (see [2, 4]) quickly established that, with the closest neighbours of the A atom being identical, the shift of the internal levels of the A atom in a compound, as compared with the element, towards higher E_b values grows in proportion to the positive state of element oxidation in that compound. With a negative oxidation state, the shift is towards smaller E_b values (negative shift). This point is extensively illustrated by the findings of studies [4, 67]. Table (1.6) cites by way of example data for platinum compounds [68, 69]. It is important to know the identity of the closest neighbours for the analysed atom. For instance, the shift of the inner levels of transition metal in complex compounds for the identical closest neighbours grows 1 eV per unit of oxidation number [4, 67], but the shift of the Pt $4f_{7/2}$ in $K_2Pt(NO_2)_4$ is higher than it is in K_2PtBr_6. This is explained by the fact that the shift ultimately depends on the charge of the platinum atom. Positive atom charges markedly grow with the growth of their oxidation state [70] but the charge also depends on the electronegativity of neighbouring atoms.

Although shift ranges for different oxidation numbers may slightly overlap depending on the character of the surrounding atoms (see Table 1.6), XPS offers today the most universal means of determining oxidation numbers. It is

Table 1.6
Correlation of binding energies Pt $4f_{7/2}$ (eV) and oxidation state Pt(N) ($L = PPh_3$)

Compound	Pt $4f_{7/2}$	N	Compound	Pt $4f_{7/2}$	N
Pt	71.3	0	$Pt(NH_3)_2(NO_2)_2$	73.9	2
L_3Pt	71.6	0	$Pt(NH_3)_2Cl_2$	73.9	2
$L_2Pt(PhC_2Ph)$	72.3	0	$[Pt(CH_3CONH)_2NO_3]_x$	75.2	3
$L_2Pt(C_2H_4)$	72.4	0	$[Pt(CH_3CONH)_2Cl]_x$	75.0	3
$[LPt(SPPh_2)]_2$	72.0	1	K_2PtF_6	77.8	4
$K_2Pt(NO_2)_4$	74.3	2	$K_2Pt(NO_2)_6$	76.1	4
K_2PtCl_4	73.1	2	K_2PtCl_6	75.7	4
$K_2Pt(SCN)_4$	72.8	2	K_2PtBr_6	74.8	4
K_2PtBr_4	72.8	2	K_2PtI_6	73.6	4
$[Pt(NH_3)_4]Cl_2$	73.6	2	$(Et_3P)_2PtCl_4$	76.1	4

*Accuracy for the photoelectron spectra of valence levels in solids and gases is markedly higher and can reach ± 0.01 eV.

Table 1.7
Binding energies Ni 2p (eV)

Ni	NiO	ΔE	Instrument type	Normalization	Reference
853.0	854.6	1.6	VIEE	C 1s (285.0)	[71–73]
852.8	856.2[a]	3.4	VIEE	Au 4f (83.0)	[74]
852.5	854.0	1.5	McPherson	Au 4f (84.0)	[75]
852.9	854.5	1.6	HP5950A	Au 4f (84.0)	[76]
853.3	854.9	1.6	AEI ES300		[77]

[a]This value seems to belong to $Ni(OH)_2$; Ni $2p_{3/2}$ in $Ni(OH)_2$ is roughly 1.5 eV higher than NiO.

nevertheless important to bear in mind that the oxidation state or number notion is formal and in many cases only has a very restricted and conventional meaning.

As the surface of a metal and or alloy is analysed, the question often arises as to whether the surface is oxidized and what component of the alloy is oxidized. X-ray photoelectron spectra can help answer this question in most cases because the binding energy of metal usually is a few eV below that of oxidized metal and the positive shift of E_b grows along with the growth of the oxidation state. E_b is also higher in hydro-oxides than it is in oxides. Regrettably, as we have shown earlier, it does not seem possible at present to recommend as standard binding energies in all the metals and their oxides despite the immensely large amount of experimental data: the scattering of data cited in literature is usually very large. Table (1.7) cites by way of example a far from complete list of published data for Ni$2p_{3/2}$ energy in metallic nickel and in NiO.

The value ΔE_b between metal and oxide does not depend on spectrometer calibration and is usually duplicated with a high degree of accuracy in different studies. Table (1.8) supplies data for binding energies in metals and oxides to give an idea of ΔE_b values between metals and oxides. It should be noted that the values given in Table (1.8) were obtained by the authors of the references cited with different energy calibration methods, so the absolute energy values may be inaccurate. The fullest available list of relatively accurate values is given in references [11, 78] and in the supplement. See also reference books [10, 14] and review [170].

E_b values for metal and oxide are close in a number of cases, and shifts in Auger-lines should be measured in such cases. Incidentally, the X-ray photoelectron spectrometer always makes it possible to obtain Auger spectra

Table 1.8
Binding energies (eV) in metals and oxides[a]

Compound	Line	E_b	Reference	Compound	Line	E_b	Reference
Li	1s	48.7	[79]	Zn	$2p_{3/2}$	1022.1	[93]
LiOH	1s	54.9	[79]	ZnO	$2p_{3/2}$	1022.5	[93]
Be	1s	~111.4	[2]	Ga	3d	18.2	[46]
BeO	1s	~114.2	[2]	Ga_2O_3	3d	20.4	[46]
B	1s	187.3	[80]	As	3d	41.6	[46]
B_2O_3	1s	191.3	[80]	As_2O_3	3d	44.6	[46]

(Table contd. over)

Table 1.8 (*Contd.*)

Compound	Line	E_b	Reference	Compound	Line	E_b	Reference
Na	1s	1071.8	[81]	Ge	$2p_{3/2}$	1117.6	[94]
	2s	63.55	[81]	GeO_2	$2p_{3/2}$	1120.5	[94]
Na_2O	1s	1072.5	[81]	Se	$3p_{3/2}$	159.9	[95]
	2s	64.0	[81]	SeO_2	$3p_{3/2}$	163.8	[95]
Mg–MgO	1s	~1	[82]	Rb	$3d_{5/2}$	112.0	[96]
Al	2p	72.6	[83]	$Rb_7Cs_{11}O_3$	$3d_{5/2}$	111.9	[96]
	2s	117.7	[83]	Sr	$3d_{5/2}$	134.2	[85]
Al	2p	75.3	[83]	SrO	$3d_{5/2}$	135.1	[85]
	2s	120.1	[83]	Zr	$3d_{5/2}$	178.5	[87]
Si	2p	99.5	[58]	ZrO_2	$3d_{5/2}$	182.6	[87]
SiO_2	2p	103.7	[58]	Nb	$3d_{5/2}$	202.4	[87]
K	$2p_{3/2}$	294.7	[84]	Nb_2O_5	$3d_{5/2}$	207.1	[87]
K_2O	$2p_{3/2}$	294.9	[84]	Mo	$3d_{5/2}$		[97][a]
Ca	$2p_{3/2}$	345.7	[85]	MoO_2	$3d_{5/2}$		[98][a]
CaO	$2p_{3/2}$	350.5	[85]	MoO_3	$3d_{5/2}$		
Sc	$2p_{5/2}$	399.0	[86]	Sc_2O_3	$2p_{3/2}$	402.8	[86]
Ti	$2p_{3/2}$	454.0	[87]	Ru	$3d_{5/2}$	279.9	[99]
TiO_2	$2p_{3/2}$	458.8	[87]	RuO_2	$3d_{5/2}$	282.1	[99]
V	$2p_{3/2}$	512.6	[88]	Rh	$3d_{5/2}$	307.1	[100]
V_2O_3	$2p_{3/2}$	515.5	[88]		$3d_{5/2}$	307.3	[101]
V_2O_5	$2p_{3/2}$	517.6	[88]	Rh_2O_3	$3d_{5/2}$	308.7	[100]
Cr	$2p_{3/2}$	573.1	[89]		$3d_{5/2}$	308.5	[101]
Cr_2O_3	$2p_{3/2}$	575.6	[89]	Pd	$3d_{5/2}$	335.3	[102]
MnO	$2p_{3/2}$	641.1	[90]	PdO	$3d_{5/2}$	337.6	[102]
Mn_3O_4	$2p_{3/2}$	641.5	[90]	Ag	$3d_{5/2}$	368.2	[93]
Mn_2O_3	$2p_{3/2}$	642.1	[90]	Ag_2O	$3d_{5/2}$	367.8	[93]
Fe	$2p_{3/2}$	706.5	[91]	Cd	$3d_{5/2}$	404.9	[93]
FeO	$2p_{3/2}$	709.6	[91]	CdO	$3d_{5/2}$	404.2	[93]
Fe_3O_4	$2p_{3/2}$	710.8	[91]	In	$3d_{5/2}$	444.2	[103]
Fe_2O_3	$2p_{3/2}$	711.6	[91]	In_2O_3	$3d_{5/2}$	445.1	[103]
Co	$2p_{3/2}$	778.2	[92]	Sn	$3d_{5/2}$	484.5	[104]
CoO	$2p_{3/2}$	780.5	[92]	SnO	$3d_{5/2}$	486.5	[104]
Co_3O_4	$2p_{3/2}$	779.6	[92]	SnO_2	$3d_{5/2}$	486.7	[104]
Ni	$2p_{3/2}$	853.0	[71–73]	Sb	$3d_{5/2}$	537.8	[104]
NiO	$2p_{3/2}$	854.6	[71–73]	Sb_2O_3	$3d_{5/2}$	539.3	[104]
Cu	$2p_{3/2}$	932.6	[93]	Sb_2O_5	$3d_{5/2}$	540.3	[104]
Cu_2O	$2p_{3/2}$	932.4	[93]	ReO_2	$4f_{7/2}$	42.5	[107]
CuO	$2p_{3/2}$	933.6	[93]	ReO_3	$4f_{7/2}$	44.5	[107]
Te	$3d_{5/2}$	572.9	[95]	Re_2O_7	$4f_{7/2}$	46.9	[107]
TeO_3	$3d_{5/2}$	576.8	[95]	Os	$4f_{7/2}$	50.6	[99]
Cs	$3d_{5/2}$	726.3	[96]	OsO_2	$4f_{7/2}$	52.7	[99]
Cs_2O	$3d_{5/2}$	725.1	[96]	Pt	$4f_{7/2}$	70.7	[108]
Ba	$3d_{5/2}$	779.1	[85]	PtO	$4f_{7/2}$	73.3	[108]
BaO	$3d_{5/2}$	778.9	[85]	Pt	$4f_{7/2}$	71.0	[101]
Ta	$4d_{5/2}$	226.9	[105]	PtO	$4f_{7/2}$	72.3	[101]
Ta_2O_5	$4d_{5/2}$	230.8	[105]	Pb	$4f_{7/2}$	137	[109]
W	$4d_{5/2}$	244.2	[105]	PbO	$4f_{7/2}$	137.9	[109]
WO_3	$4d_{5/2}$	248.3	[105]	PbO_2	$4f_{7/2}$	137.5	[109]
W	$4f_{7/2}$	31.5	[106]	Bi	$4f_{7/2}$	157.1	[110]
WO_2	$4f_{7/2}$	34.6	[106]	Bi_2O_3	$4f_{7/2}$	159.5	[110]
WO_3	$4f_{7/2}$	36.1	[106]	U	$4f_{7/2}$	377.0	[111]
Re	$4f_{7/2}$	40.7	[107]	UO_2	$4f_{7/2}$	380.3	[111]
				Pu	$4f_{7/2}$	422.2	[112]
				Pu_2O_3	$4f_{7/2}$	424.4	[112]
				PuO_2	$4f_{7/2}$	426.1	[112]

[a]See discussion..

Table 1.9
Binding energies (eV)

Oxide	O 1s	M nl	*k*	Oxide	O 1s	M nl	*k*
H_2O	533.5			ZrO_2	530.4	182.4	1.26
BeO	531.8	113.9	1.35	Nb_2O_5	530.8	207.8	1.28
B_2O_3	533.2	193.5	1.32	MoO_3	531.1	233.3	1.10
MgO	530.2	88.3	1.32	CdO	529.4	404.4	1.71
Al_2O_3	531.8	75.0	1.53	In_2O_3	530.0	444.5	1.20
SiO_2 (a)	533.1	103.8	1.58	SnO_2	530.9	486.9	1.30
SiO_2 (b)	533.1	104.0	1.31	La_2O_3	529.0	101.5	
	532.4			Ce_2O_3	529.4	882.4	—
P_2O_5		135.4		Nd_2O_3	529.4	121.5	
	533.8			Sm_2O_3	529.2	131.6	—
Sc_2O_3	530.3	402.1	1.55	Eu_2O_3	529.2	135.1	—
TiO_2 (a)	530.8	459.2	1.46	Gd_2O_3	529.3	141.6	—
TiO_2 (b)	531.3	459.1	1.23	Dy_2O_3	529.4	156.0	—
TiO_2 (c)	531.0	459.1	1.78	Ho_2O_3	529.4	161.5	—
TiO_2 (d)	529.4	458.8	1.90	Er_2O_3	529.5	168.4	—
V_2O_5	530.8	517.9	1.20	Tm_2O_3	529.6	176.0	—
Cr_2O_3	530.7	576.1	1.05	Yb_2O_3	529.5	184.5	—
CrO_3	530.8	579.7	0.85	Lu_2O_3	529.7	196.2	—
MnO_2	530.2	642.8	—	HfO_2	530.6	213.5	—
Fe_2O_3	530.3	711.0	1.66	Ta_2O_5	530.8	27.1	—
CoO	530.4	780.6	1.16	WO_3	530.8	36.2	1.09
Co_3O_4	530.4	780.5	1.15	PbO (?)	529.1	137.9	—
NiO	529.9	854.6	1.13	PbO_2 (?)	529.7	137.9	
Cu_2O (?)	530.9	932.8	—	Bi_2O_3	530.0	159.0	—
CuO	530.9	934.3	—	ThO_2	530.2	86.0	—
ZnO	530.4	1021.7	—				
Ga_2O_3	530.9	105.8	—				
GeO_2	532.4	125.4	—				
Y_2O_3	529.5	157.1	—				

N.B. M nl is the most intensive line in metal; *k* is the indicator in stoichiometric formulae of the SiO_{2k} and Al_2O_{3k} type, determined on the basis of X-ray photoelectron analysis.

excited by X-ray lines. E_b values for oxides, determined in studies [71, 72] and referenced to the C 1s line (285.0 eV) can also be used to determine the oxidation state of metal. (Table 1.9).

1.3.2. *Dependence of the* E_b *value on neighbouring atoms and its additive character*

If oxidation numbers are equal, the positive shift of E_b in the atom under analysis increases with the growth of neighbouring atom electronegativity. This was established in the early works on XPS [2, 3] and effectively confirmed by later experimental findings [4].

Let us study in some detail the dependence of E_b on the character of the closest environment. An analysis of a vast body of experimental data shows [4, 113] that in transition elements the ability of acidoligands for an increase in E_b of the transition element is represented by the series

$$\mathrm{F} > \mathrm{NO_3} \gtrsim \mathrm{NO_2} \gtrsim \mathrm{CH} \sim \tfrac{1}{2}\mathrm{C_2O_4} \sim \mathrm{NCS} > \mathrm{Cl} > \mathrm{Br} \sim \mathrm{SCN} > \mathrm{CH_3}, \quad \mathrm{Ph}$$

The following series have been evolved for non-transition elements and they help identify the character of the chemical compound being analysed [114]:

$$\text{Pb: } NO_3 \sim \tfrac{1}{2}SO_4 > F \gtrsim Cl \sim \tfrac{1}{2}C_2O_4 \sim CH_3COO \sim NCO \gtrsim Br \sim \tfrac{1}{3}PO_4 \gtrsim NCS \sim I > \tfrac{1}{2}O;$$

$$\text{Cd: } ClO_4 > Cl > Br > I > CH_3COO > \tfrac{1}{2}O;$$

$$\text{Bi: } F > Cl > Br > \tfrac{1}{2}O \gtrsim I;$$

$$\text{Sb: } F > Br > \tfrac{1}{2}O > I > \tfrac{1}{2}S > Ph.$$

The analysis of data shows that in principle the experimental values ΔE_b of atom A can be represented with sufficient accuracy as the total $\sum(j)\Delta E(A - B_j)$, where $\Delta E(A - B_j)$ is the contribution of the $A - B_j$ bond to the total shift. The limited applicability of different additive patterns is well known. Here the compilation of the additive pattern is justified, first by the opportunity it provides to evalute the energy shift of the inner atomic level in different surroundings, which is of interest to the interpretation of the spectra of compounds containing several non-equivalent atoms of that element and, second, because it allows the quantitative study of a number of regularities in greater detail.

The additive patterns of shifts have now been compiled for the compounds of many elements, such as C [115, 116], N [117], Si [118], P [117–119], S [115], Pt, Pd, Ir, Rh and Co [4, 113], and for some transition elements [120].

We will cite by way of an example data for the calculation of C 1s energies (Table 1.10 [115]). The data of Table (1.10) make it possible to calculate the C 1s shift for virtually any aliphatic compound. The zero value E_0 equals 285 eV, so absolute values also are easy to obtain. The correspondence of the calculated and experimental values is quite good. For instance, the shifts of the calculated energies C 1s relative to $E_0 = 285$ eV for C atoms in the $CH_3CH_2OC(O)CH_3$ left to right are 0.6; 2.2; 4.5; and 0.3 eV, while the experimental values are 0.6; 2.1, 4.4 and 0.6 eV; these values in $CH_3C(O)CF_3$ are 0.3; 3.8; 8.4 and 0.4; 3.4; 7.8, respectively. Additive patterns are therefore quite useful in identifying chemical compounds.

Reference [121] suggests a somewhat more complex additive pattern for the calculation of ΔE. The value ΔE is presented as $\Delta E = aF + bR$ where a and b depend on the type of the molecule and the atom being analysed, while F and R reflect the σ- and π-electronegativity of substitutes.

Table 1.10
Value $\Delta E(C{-}B_j)$ for the C 1s line (eV)

B_j	ΔE	B_j	ΔE	B_j	ΔE
H^a	0	$=O$	3.0	$-CH_kF_m$	$0.3\,m$
$CH_nR_m{}^b$	−0.1	$-NH_2$	0.8	$-CH_k(NH_2)_m$	0.0
$-O^-$ or OM^c	0.6	$-Br$	1.1	$-CH_k(OR)_m$	$0.2\,m$
$-OR$	1.6	$-Cl$	1.5	$-Ch_kCl_m$	$0.2\,m$
$-OC(O)Y^d$	2.3	$-F$	2.7	$CH_kOC(O)Y_m$	$0.6\,m$
$=S$	1.1	—	—	$CH_kC(O)Y_m$	$0.3\,m$

[a]By definition. [b]R is the hydrocarbon radical or H. [c]M is metal. [d]$Y = Cl$, CF_3, CH_3, OR.

1.3.3. *Constancy of* E_b *in functional groups. Identification of compounds*
The value E_b in functional groups depends very little on the chemical compound of which that functional group is a part. Figure (1.8) [122] presents data for the S 2p line in different compounds. When there are several compounds with a given functional group, the S 2p values do not usually differ by more than 1 eV. A similar conclusion can be drawn also for the N 1s line (see Fig. 1.2).

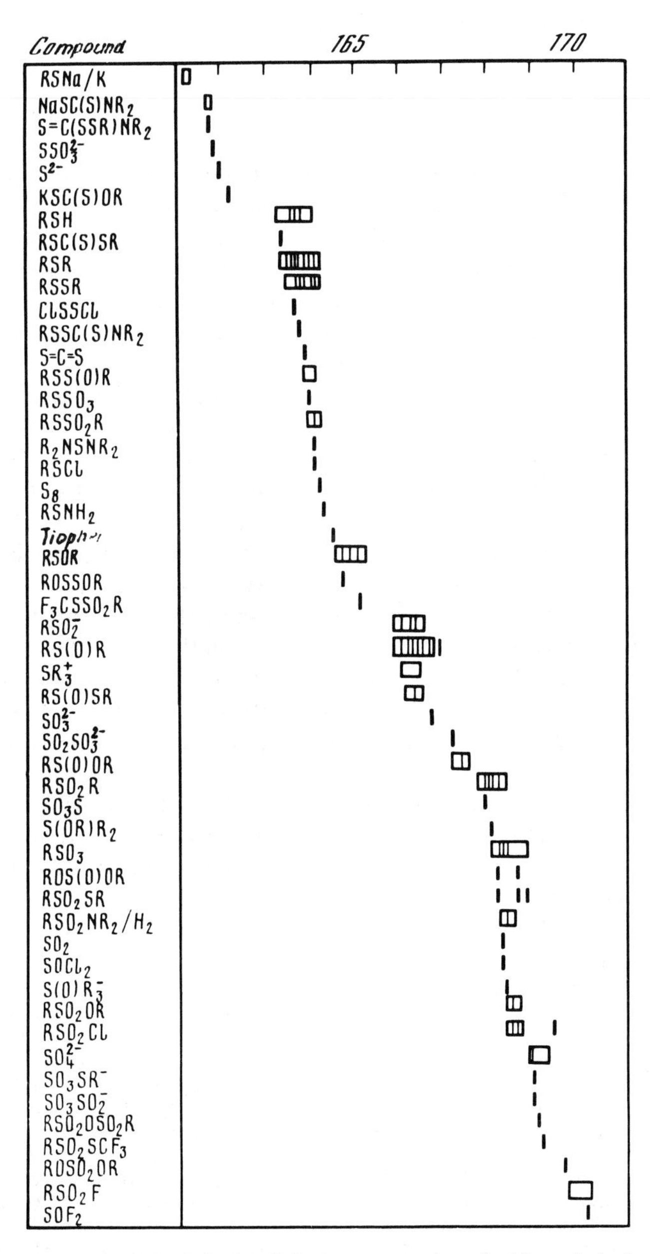

Figure 1.8. Binding energy S 2p. R is the aliphatic or aromatic radical bound via the C atom with the S atom (top compound).

Table 1.11
Binding energies (eV)

Group	C 1s	Group	N 1s	Group	A 2p
—C(O)OR	289.5	NO_3^-	408–407	SO_4^{2-}	169–170
—COOM	289.0	NO_2^-	404.6–403.6	SO_3^{2-}	167–168
R_3COH	286.7	RNO_2	406.0	Sulphides	162–163
R_2CO	288.0	NH_4^+	401–402	PO_4^{3-}	133
CO_3^{2-}	289–290	NH_3	400.4	Phosphides	128–130
Carbides	282–283	NCS^-	398.5	ClO_4^-	208–209
Silicides (Si 2p)	99–100	CN^-	399.0	ClO_3^-	206–207
Borides (B 1s)	188–189	Nitrides	397–398	Chlorides	198–200

Table (1.11) cites typical binding energy values for some major functional groups (C 1s = 285 eV).

The binding energy within a given functional group regularly varies depending on the structural characteristics of the compound.

1. In the case of anions, the binding energy of their inner electrons grows with a rise in the electronegativity of the cations, with the largest shift occurring in the transition to the proton. For instance, S 2p = 168.9; 169.4; 169.8 eV in K_2SO_4, Li_2SO_4, and $KHSO_4$; C 1s = 289.3; 289.9; 290.0; 290.2 in K_2CO_3, Li_2CO_3, $KHCO_3$, and $NaHCO_3$ [123].

2. The binding energy of the inner electrons of the anion directly connected with metal is usually higher than it is in the anion positioned in the outer sphere of the compound. For instance, the N 1s energy of the NO_3 group equals 407.5 and 407.3 eV in $K_3Rh(NO_3)_6$ and $Rh(NH_3)_3(NO_3)_3$, and 407.1 eV in $[Rh(NH_3)_5X]NO_3$ [4].

These results also hold for simple anions of the F, Cl, O, etc. types. In particular, the binding energy F 1s regularly varies depending on the position of the cation in the Periodic Table (Table 1.12) [124]. Review [125] supplies a detailed analysis of different factors influencing the Cl 2p, Br 3d and F 1s binding energies.

O 1s values in different oxides are given in Table (1.9) [71, 72] (see also [126]). It should be noted that the O 1s line in oxides of transition elements is often accompanied by a shoulder having a higher energy (by roughly 2 eV). This shoulder is due to O atoms in OH^- groups or in H_2O molecules on oxide surface. O 1s values in the oxides of non-transition elements in the highest oxidation state depend on the position of the metal involved in the Periodic Table. The presence of OH^- groups of the adsorbed H_2O and O_2 molecules on the oxide surface usually leads to a surplus of oxygen atoms on the surface as compared with bulk. The K coefficient in Table (1.9) shows the ratio of oxygen agents on the surface to the stoichiometry of the specimens studied. This ratio depends markedly on specimen prehistory and processing (cf. different results for SiO_2 and TiO_2). The k value is determined by the X-ray photoelectron method (see Chapter 2).

The presence of OH^- groups and H_2O molecules on the oxide surface can in principle result in three peaks for O 1s. To give an idea of their relative position, we will cite data for energies in the Cu + O_2 system [127]: O 1s energies equalling 530.5, and 533.5 eV are those of the oxide OH^- groups and adsorbed H_2O (see Section 4.1 below).

Table 1.12
Binding energies F 1s (eV)

Period	Group						
	I		II	III		IV	V
2	LiF 685.2		BeF_2 686.1			K_2SiF_6 686.8	
3	NaF 684.7		MgF_2 685.9	$AlF_3 \cdot 3H_2O$ 686.5	K_2TiF_6 685.1		
4	KF 684.1		CaF_2 685.1			K_2GeF_6 685.4	
		ZnF_2 685.3		$GaF_3 \cdot 3H_2O$ 685.4	K_2ZrF_6 684.8		K_2NbF_7 685.4
5	RbF 683.1		SrF_2 684.7			$K_2SnF_6 \cdot H_2O$ 685.3	
		CdF_2 685.0		$InF_3 \cdot 3H_2O$ 685.5			K_2TaF_7 685.3
6			BaF_2 684.5				

Table 1.13
Binding energies (eV)

Compound	M nl	S 2p	Compound	M nl	S 2p
$FeS_{1.1}$	711.3	162.1	ZrS_2	183.0	161.4
	708.2		$NbS_{1.6}$	203.9	162.1
FeS_2	707.6	163.1	MoS_2	229.6	162.8
Cu_2S	932.6	162.3	Au_2S		163.3
CuS	935.4	162.1	Ag_2S		162.7
	932.7		CdS		162.0
ZnS	1022.0	162.1	Ga_2S_3		162.4
PbS	137.9	161.1	In_2S_3		162.2
HgS	101.0	162.2			

A selection of E_b values measured in our laboratory (C 1s = 285 eV) for different sulphides is given in Table (1.13) (data for Zr, Nb, No sulphides are cited in reference [128]). The S 2p energy in elementary sulphur S_8 equals 164.2 eV. Surface oxidation often leaves traces of oxides on the surface of sulphides (see data for FeS on the growth of M nl energies) and SO_4^{2-} (by maximum S 2p around 168–169 eV). A large group of borides have been studied in [129]. The M nl values in borides differ relatively little from appropriate M nl values in metals. The oxidized form B is usually present on the surface of borides.

The identification of chemical compounds by chemical shifts in X-ray photoelectron lines can be made substantially easier by a procedure suggested in [130]. That paper considers two-dimensional charts (Fig. 1.9) that show the kinetic energies of Auger-lines, with the sum total of the kinetic energy of Auger-lines and binding energies as ordinates (Y-lines) and the binding energies of inner electron as abscissae (X-lines). Since in the general case there are no simple relationships between these values, the closeness (on the X line) of the values of X-

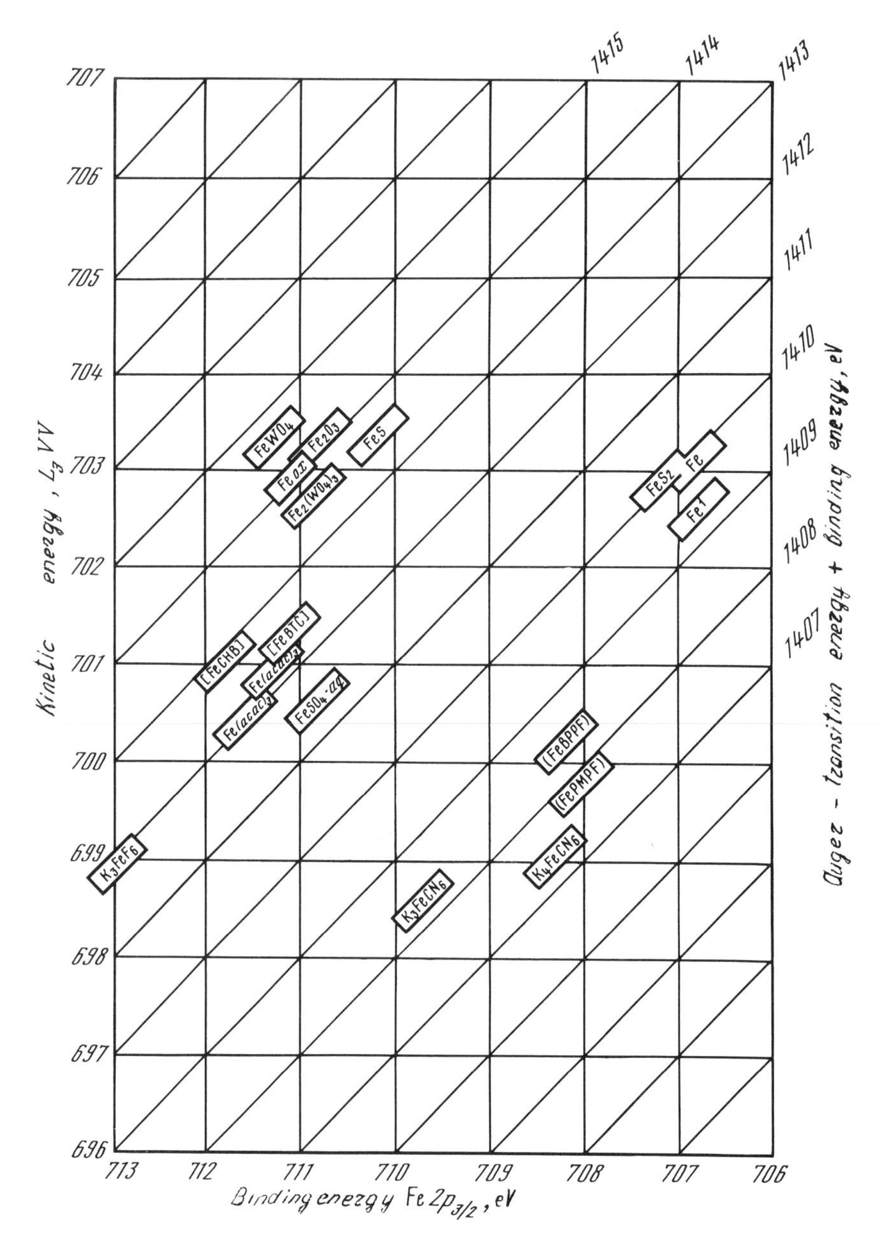

Figure 1.9. Two-dimensional representation of X-ray photoelectron and Auger chemical compound identification data for ferrous compounds.

ray photoelectron characteristics for two compounds under consideration does not necessarily indicate a close relation between the Auger-spectrum characteristics of these compounds (on the *Y* line). Hence there are additional possibilities for the identification of a chemical compound. Maps of this type have been compiled for compounds of 24 elements [130].

Extensive lists of experimental data for shifts of X-ray photoelectron lines are supplied in [10, 14]. The *Journal of Analytical Chemistry* [131–137] systematically carries bibliographies of X-ray photoelectron studies. These data are useful in sorting out experimental E_b values. In view of the relatively low accuracy of the absolute values E_b, which makes itself manifest in the large scattering of values obtained by different authors, it is useful in identifying chemical compounds on the basis of E_b or ΔE_b to study standard compounds in the same experimental conditions as the analysed substance. Recourse to published data is only possible after a thorough verification of the results obtained from the given spectrometers.

1.4. Relationship of E_b to effective atomic charge and other characteristics

1.4.1. *Atomic charge*

Let us consider the values on which the ionization energy E of an inner level in a compound depends. This is best done by the MOLCAO method:

$$E_i(\mathrm{A}) = E_{i\mathrm{A}} + E^1_{i\mathrm{kin}} + \sum(i'_\mathrm{A})E_{ii'_\mathrm{A}} + \sum(v_\mathrm{A})E_{iv} \tag{1.15}$$

A similar expression for $E_i(\mathrm{M})$ can be written for compound M, to which atom A belongs. This expression for inner level energy $E_i(\mathrm{M})$ can be divided into two parts. One will include the energy of interaction of the inner electron with the nucleus of its atom, the other inner electrons of its atom and that part of valence electrons MO the wave function of which is described by the orbitals of the considered atom, and also the kinetic energy of the given electron. In particular, the interaction of the electron i with the molecular orbital:

$$\varphi_{\mathrm{MO}} = \sum(v)(C_{v\mathrm{A}}\varphi_{v\mathrm{A}} + C_{v\mathrm{X}}\varphi_{v\mathrm{X}})$$

is broken down into two:

$$\langle \psi_i^2 | e^2/r_{i\mathrm{MO}} | \psi^2_{\mathrm{MO}} \rangle \approx \langle \psi_i^2 | e^2/r_{i,v\mathrm{A}} | C^2_{v\mathrm{A}} \varphi^2_{v\mathrm{A}} \rangle + \langle \psi_i^2 | e^2/r_{i,v\mathrm{X}} | C^2_{v\mathrm{X}} \varphi^2_{v\mathrm{X}} \rangle, \tag{1.16}$$

and only the first item is taken account in the expression $\sum_{v_{\mathrm{A(M)}}} E_{iv}$ in Equation (1.17).

The other part of the expression for the energy of the inner electron will include the expressions representing the interaction of the given inner electron with all the other electrons and nuclei with the exception of the electrons and nucleus of atom A.

That part will be denoted as Madelung energy E_{Mad}:

$$E_i(\mathrm{M}) = [E_{i\mathrm{A(M)}} + E_{i\mathrm{kin(M)}} + \sum(i'_{\mathrm{A(M)}})E_{i,i'\mathrm{A(M)}} + \sum(v_{\mathrm{A(M)}})E_{i,v}] + E_{\mathrm{Mad}} \tag{1.17}$$

Since the wave function of the inner electron differs from zero only when it is in the immediate proximity of its nucleus and since its overlap with the electron functions of other atoms is infinitesimally small, the function of the inner electron of atom A virtually remains unchanged with the transition from the atom to the compound. Hence

$$E_{i\mathrm{A}} \approx E_{i\mathrm{A(M)}}, E_{i\mathrm{kin(M)}} \approx E_{i\mathrm{kin}}, E_{i,i'\mathrm{A}} \approx E_{i,i'\mathrm{A(M)}} \tag{1.18}$$

From (1.15)–(1.18), we obtain

$$\Delta E_b = E_i(M) - E_i(A) = \sum(v_{A(M)})E_{i,v} - \sum(v_A)E_{iv} + E_{Mad} \quad (1.19)$$

The expression $\sum(v_A)E_{iv}$ can be written as

$$\sum(v_A)E_{iv} = -n_{vA}\langle\psi_i^2|C^2/r_{i,vA}|\psi_{vA}^2\rangle \approx -n_{vA}e^2\langle 1/r_{vA}\rangle, \quad (1.20)$$

where n_{vA} is the number of valence electrons in atom A; and $\langle 1/r_{vA}\rangle$ the mean value of $1/r_{vA}$.

The last equation in (1.20) is explained by the fact that ϕ differs from zero only near the nucleus of atom A, and therefore

$$\langle\psi_i^2|e^2/r_{i,vA}|\psi_{vA}^2\rangle \approx \langle e^2/r_{A,vA}\psi_{vA}^2\rangle \equiv e^2\langle 1/r_{vA}\rangle. \quad (1.21)$$

Since $\sum(v_{A(M)})C_{vA}^2 = n_{vA}(M)$, where $n_{vA}(M)$ is the number of valence electrons of atom A in compound M, we have, in view of (1.16), (1.19), and (1.20):

$$\Delta E_b = q_{A(M)}\langle e/r_{vA}\rangle + E_{Mad}, \quad (1.22)$$

were $q_{A(M)} = n_{vA} - n_{vA}(M)$, the effective charge of atom A in M.

Similar to (1.21), the interaction of electron i of atom A with the nuclei and electrons of atoms X of molecule M can be represented as $E_{Mad} = q_X/r_{A,X}$, where q_X is the effective charge of atom X. We finally obtain

$$\Delta E_b = Kq_A + \sum(X)q_X/r_{A,X}, \quad (1.23)$$

It should be noted that q_A and q_X usually have different signs, so E_b constitutes a small difference between large values.

Equation (1.23) can also be obtained from qualitative model considerations [2]. The binding energy of the inner electron depends on the potential created by all the electrons. Since the wave functions of only valence electrons change in a transition from the free atom to the atom in a compound, it is enough to limit our analysis to them. Let the average value of the potential created in the atom by the valence electron equal $1/r$, where r is the distance from the electron to the nucleus of the given atom. If that electron in a compound is completely transferred to other atoms and the mean distance from that electron to the nucleus is R_{AX}, the change of the potential will equal $1/r - \sum(X)\, 1/R_{AX}$. We have for effective charge q_A on the given atom A $\Delta E_b = q_A(1/r) + \sum(X)q_X/R_{AX}$, which is equivalent to (1.23).

An expression like (1.23) was obtained or used in references [2, 3, 138, 139] and others on the basis of arguments similar to those given above.

Many studies (see bibliography in [4, 14]) show that E_b correlates with the effective charge of the given atom. The atomic charge was calculated with the help of Pauling electronegativities, the Extended Huckel Method, MO methods in CNDO (complete neglect of differential overlap), and SCF (self-consistent field) approximations. The dependence of E_b on q is usually almost linear (Fig. 1.10 [140]).

The existence of a correlation between q and E_b is not a trivial fact. Calculations show that E_{Mad} in (1.23) for free molecules and crystals is usually several times higher than ΔE_b because the latter is a small difference between two

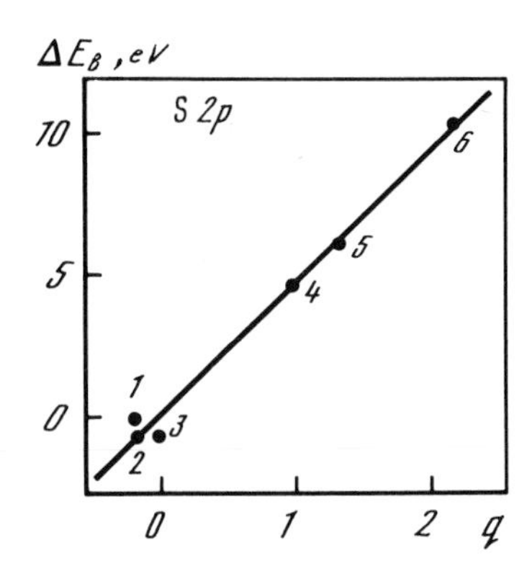

Figure 1.10. Dependence of the S 2p line shift on the charge q of the S atom. $1 = H_2S$; $2 = CS_2$; $3 = C_4H_4S$; $4 = SO_2$; $5 = SO_2F_2$; $6 = SF_6$.

large values. Usually, the correlation of ΔE_b with the computed $kq \pm E_{Mad}$ is noticeably higher than with q (see bibliography in [4]). The correlation of ΔE_b and q, computed by different methods, can obviously be used to evaluate q (within the bounds of the given method). The latter reservation is quite important. For instance, E_b of the 1s-electron of carbon correlates both with q_p charges according to Pauling [141], q_h [141] and q_c [3, 116] charges obtained by the Extended Huckel Method and CNDO, though the values of the charges are noticeably different from one method to another. For instance, the charges of atom C in CF_4 equal: $q_p = 1.72$, $q_h = 2.83$, $q_c = 0.85$. The evaluations of q_A on the basis of E_b are considered in [4, 142–144].

The simbatic progress of the effective charge and E_b is always observed only with large differences in ΔE_b (or q). It is quite possible that failures to follow the simbatic progress with small differences in q are explained by inaccuracies in the evaluation (or computation) of the effective charge. However, in view of the substantial influence of E_{Mad} and E_{rel} (see below) on the shift, the absence of correlation in some cases does not seem inappropriate.

The above approach to the relationship of ΔE_b and q is known as the electrostatic model of the chemical shift, or the model of the ground state potential. Usually, Equation (1.23) is used for analysis in the following form:

$$E_b = kq + E_{Mad} + l \qquad (1.24)$$

Value l is introduced to take account of the starting point for calculating ΔE_b and also relaxation correction (see below). The values k and l were determined in different studies (the depend on the method of determing q). Table (1.14) [145] gives values of k and l for some X-ray photo-electron lines, with ΔE_b computed for the C 1s, N 1s, and O 1s lines and the absolute value E_b determined for the other lines.

In most cases the change of sign of value ΔE_b in a transition from one chemical compound to another is determined by the sign of q, though in some instances the variation of E_{Mad} plays the decisive role. There also are cases (see for example [146, 147]) where the variation of relaxation energy E_{rel} is most important.

Table 1.14
k and *l* values

Line	k	l	Line	k	l
C 1s	29.89	8.04	P 2p	19.28	139.37
N 1s	30.23	5.22	S 2p	18.88	172.35
O 1s	26.61	2.61	Cl 2p	18.27	208.34
F 1s	28.68	91.98	Ge 3p	15.87	130.89
Si 2p	17.29	110.07	Br 3d	13.40	76.96

Relaxation energy is the energy related to the change of wave function caused by the ejection of an electron from the atom in photoionization. The ionization energy ε is computed without taking into account the relaxation of wave functions*, while ε_{rel} is the ionization energy taking account of the wave functions. Then

$$E_{rel} = \varepsilon_{rel} - \varepsilon. \tag{1.25}$$

Relaxation can be explained as redistribution of electronic density within a given atom (inter-atomic relaxation) and between atoms (extra-atomic relaxation). Inter-atomic relaxation can reach dozens of electron-volts and extra atomic relaxation energy up to 10 eV [147–150]. However, E_{rel} for a given atom changes little from one compound to another and for this reason a substantial role is usually played by kq, while l is considered constant in (1.24). Relaxation energy plays a significant role in determining E_b in analysing differences between E_b in the atomic gas–solid transition [148–150] (because in this case $\Delta q = 0$), and also in analysing differences between E_b and E_{ion} in free and adsorbed molecule (see [150, 154] and Chapter 4).

References [152, 153] consider a model which takes account of relaxation energy but which is for the most part close to the ground state potential model (1.24). An equation coinciding with (1.24) was obtained by an analysis of transition state T, which corresponds to an electron half removed from the considered level i of atom A; however, values k, q, E_{Mad}, and l in it belong to the transition rather than to the ground state. It is known as the transition state model.

The more consistent approach to calculating ΔE_b presupposes an analysis of the initial (non-ionized) and final (ionized) states of compounds:

$$E_b = E_{in} - E_{fin}, \tag{1.26}$$

where E_{fin} and E_{in} are total energies of the respective states.

When E_b is presented as in (1.26), the value E_b and the energies of the states are positive values, as was assumed above. Hence,

$$\Delta E_b = \Delta E_{in} - \Delta E_{fin}. \tag{1.27}$$

It was demonstrated above [see (1.23)] that:

$$\Delta E_{in} = k\Delta q + \Delta E_{Mad} \approx \Delta V.$$

*This approach is based on Koopmans' theorem (for details see [5, 12]).

Table 1.15
ΔE_b, ΔV and ΔR values (eV)

Compound	ΔR	ΔV	ΔE	Compound	ΔR	ΔV	ΔE
i-C_4H_9Cl	0.63	−0.25	−0.88	$CHCl_3$	0.81	1.41	0.60
n-C_3H_7Cl	0.36	−0.08	−0.44	CCl_4	0.99	1.77	0.78
C_2H_5Cl	0.31	−0.03	−0.34	HCl	−0.71	0.41	1.12
CH_3Cl	0	0	0	Cl_2	0.39	1.95	1.56
CH_2Cl_2	0.50	0.86	0.36	ClF	−0.43	2.49	2.93

This is the change of potential V near the nucleus of the given atom for the initial states of the studied and standard compounds. References [154, 155] demonstrated that often $\Delta V \approx E_b$.

The value ΔE_{fin} includes the difference of relaxation energies and possible geometric changes in the ionized state as compared with the initial state. Let us assume that $\Delta E_{fin} = \Delta R$.

ΔE_b can therefore be represented as the difference of ΔV and ΔR, where ΔV corresponds to the change of the electrostatic potential near the nucleus for the initial state of the molecule, and R to the change of relaxation energy or, to be more precise, to the change of the electrostatic potential near the nucleus for the final state:

$$\Delta E_b = \Delta V - \Delta R. \tag{1.28}$$

Using similar arguments, the value of the shift ΔK for the kinetic energies of the Auger-transition can be represented (see bibliography in [156, 157]), as

$$\Delta K = -\Delta V + 3\Delta R. \tag{1.29}$$

Hence:

$$\Delta K + \Delta E_b = 2\Delta R. \tag{1.30}$$

Equation (1.30) can be used for the experimental evaluation of the contribution of V and R to ΔE_b. There evaluations are cited in [156] for some Cl, Ge, and P compounds and in [157] for Si-compounds. Table (1.15) supplies data for a number of compounds of Cl, with values for CH_3Cl taken as zero. These problems are discussed in greater detail in [14].

1.4.2. *Thermodynamic characteristics*

Jolly [158, 159] established a very interesting relationship between shifts of X-ray photoelectron lines and energies of some reactions.

The value $E_{b(ion)}$ is energy of processes

$$M \rightarrow M^+(1s) + e, \tag{1.31}$$

where M is the molecule and the value 1s in brackets shows that, for example, the 1s-electron in one of the atoms of the molecule has been ejected. Jolly presumed that as a result of electron relaxation, molecule $M^+(1s)$ transforms into another molecule, in which the atom (number Z) with the ejected 1s-electron is replaced by atom $Z+1$, for instance, $CH_4^+(1s) \rightarrow NH_4^+$. This proposition is well supported by X-ray adsorption spectra [70], and also by computations of molecule $O_2^+(1s)$ [160], done already after Jolly's studies. The computation [160] demonstrated that the population of the valence orbitals in $O_2^+(1s)$ is indeed close to the population of the FO^+ orbitals. It therefore can be expected that energy E of the transition, e.g., $CH_4 \rightarrow NH_4^+$ is roughly equivalent to the value C 1s in CH_4. However, the following equation is more accurate

$$\Delta E = \Delta E_b, \tag{1.32}$$

where ΔE and ΔE_b are appropriate differences for two similar processes, such as $CH_4 \rightarrow NH_4^+$ and $CO_2 \rightarrow NO_2^+$. It suffices to presume for (1.32) to hold that the energy of all reactions of the type

$$CH_4^+(1s) + \tfrac{1}{2}N_2 \rightarrow NH_4^+ + C \quad \text{(graphite)}$$
$$CO_2^+(1s) + \tfrac{1}{2}N_2 \rightarrow NO_2^+ + C \quad \text{(graphite)}$$

is roughly equal. The correctness of (1.32) has been corroborated by extensive materials (Fig. 1.11 [159]). At present [161] a relationship has been established between thermodynamic values and shifts in the lines C 1s, B 1s, N 1s, O 1s, P 2p and S 2p. This approach has come to be known as the method of equivalent cores. The relationship of this model with the electrostatic one is analysed in [162]. The method of equivalent cores is also useful in computing values of ΔE_b in atoms in solids and gases (see Section 4.7), the difference E_{sb} for atoms on the surface and in the bulk of solids (see Section 5.2), and to evaluate shifts in lines following the adsorption of atoms and molecules on the surface of solids [150, 163]. Let us consider this last point in greater detail [150].

Let us write down total energy E_{MZ} of the system consisting of matrix M and

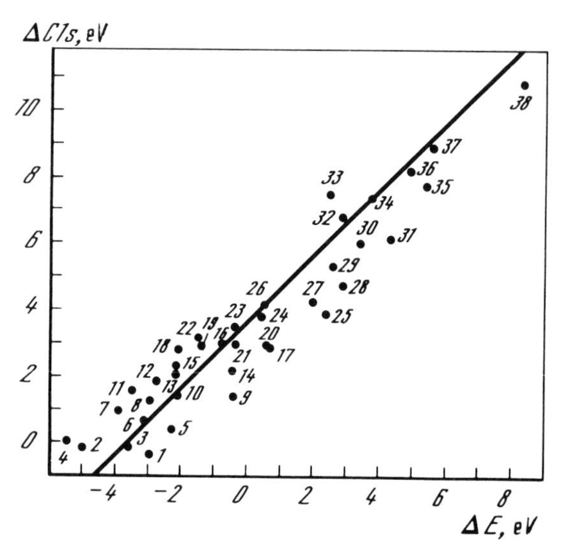

Figure 1.11. Relationship between the C 1s line shift and reaction energy ΔE. (See [159] for names of substances indicated by figures along the line.)

adsorbed A atoms* with ordinal number Z:

$$E_{MZ} = E_M + E_Z + E_{(MZ)}, \tag{1.33}$$

where E_M and E_Z are the energies of free M and A; $E_{(MZ)}$ is the energy of interaction.

Let an inner electron be ejected from atom A. The energy of the system may be written in a similar manner, where '+' indicates the removal of the electron:

$$E_{MZ^+} = E_M + E_{Z^+} + E_{(MZ^+)}. \tag{1.34}$$

Subtracting (1.33) from (1.34), we obtain

$$E_{MZ^+} - E_{MZ} = E_{Z^+} + E_{(MZ^+)} - E_Z - E_{(MZ)}. \tag{1.35}$$

Denoting the ionization potentials of the adsorbed and free atom A as I_{MZ} and I_Z, we obtain

$$I_{MZ} = I_Z + E_{(MZ^+)} - E_{(MZ)}. \tag{1.36}$$

The value $I_Z - I_{MZ}$ is ΔE_b for the adsorbed and free atom A, therefore

$$\Delta E_b = E_{(MZ)} - E_{(MZ^+)}. \tag{1.37}$$

Proceeding from the method of equivalent cores, let us substitute for $E_{(MZ^+)}$ the value $E^{ion}_{(M(Z+1))}$, which corresponds to the interaction of A atom with the ordinal number Z + 1 without one valence electron with matrix M. Value $E^{ion}_{(M(Z+1))}$ may be approximately represented as

$$E^{ion}_{(M(Z+1))} = E_{(M(Z+1))} - I_{Z+1} + \varphi, \tag{1.38}$$

where φ is work function of matrix M. Equation (1.38) takes account of the fact that transition from $E_{(M(Z+1))}$ to $E_{(M(Z+1))}$ requires energy inputs for ionizing A atom, which is partially compensated for by the work function of matrix M.

In view of (1.37) and (1.38), we obtain

$$\Delta E_b = E_{(MZ)} + I_{(Z+1)} - \varphi - E_{(M(Z+1))}. \tag{1.39}$$

Table 1.16
Parameters of equation (1.39) (eV)

System	$E_{(M(Z+1))}$	E_{MZ}	$I_{(Z+1)}$	φ	E_b (exp)	E_b^* (1.39)
C/W	6.0	10.3	14.5	4.6	8.2	5.6
N/W	5.2	6.0	13.6	4.6	7.6	8.2
O/W	3.5	5.2	17.4	4.6	9.9	11.1
CO/W	0	0	9.3	4.6	5.9	4.7
CO/Ni(100)	1.1	1.3	9.3	5	5.1	4.1
CO/Ni(111)	1.1	1.1	9.3	5	5.5	4.3
CO/Cu	1	1	9.3	4.5	5.1	4.8
CO/Pt(100)	1.2	1.4	9.3	5.8	4.7	3.3
CO/Pt	1.2	1.4	9.3	5.6	4.0	3.5
O/graphite(0001)	7.1	13.6	17.4	5	8	6.0

*For the C, N, and O 1s lines in free atoms and the C 1s line in CO.

*If we consider an adsorbed molecule, atom A corresponds to the analysed atom in the molecule.

Equation (1.39) can be used to calculate ΔE_b of adsorbed and free atoms and molecules. For instance, in N atom, $E_{(MZ)}$ and $E_{(M(Z+1))}$ for the N 1s line correspond to the adsorption heat of N and O atoms, and in CO molecule in calculating ΔE_b for the C 1s line, $E_{(MZ)}$ and $E_{(M(Z+1))}$ correspond to the adsorption heat of CO and NO, and I_{Z+1} is the ionization potential of No molecule. Table (1.16) [150] supplies relevant values and draws a comparison between the computed and experimental values of ΔE_b.

1.4.3. *Inter-atom distances**

Let us consider changes in parameters q, E_{mad} and l in Equation (1.24) depending on the distance R_{AB} between atom A and the surrounding B atoms. Member kq (in absolute value) stands in direct proportion to q, i.e. in reverse proportion to the electronegativity of metal or metalloid B, with which negatively charged atom A is bound. In the first approximation, electronegativity stands in reverse proportion to the size of atom B. Expression $\sum(B)q_B/R_{AB}$ is in reverse proportion to R_{AB}. Relaxation energy (l) contributes to lessening the binding energy of the inner levels of A element, and we may presume that the quantity of electron density transported from B atom to A atom in the photoionization of A atom is in proportion to the polarization of B atom, i.e. to its size.

Therefore, all the three components on which the E_b shift depends under Equation (1.24) contribute to lessening the binding energy of inner electrons in negatively charged A atom in the compound as the average distance A–B increases:

$$E \sim 1/R_{AB}. \tag{1.40}$$

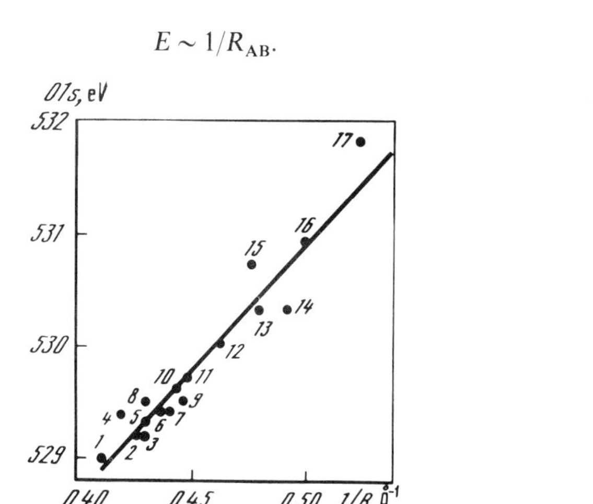

Figure 1.12. Correlations between the O 1s energy and the $1/R$ parameter in M_2O_3 oxides. 1 = La; 2 = Sm; 3 = Eu; 4 = Nd; 5 = Gd; 6 = Dy; 7 = Ho; 8 = I; 9 = Y; 10 = Tm; 11 = Ca; 12 = In; 13 = Sc; 14 = Fe; 15 = Cr; 16 = Ga; 17 = Al.

*See references [4, 154, 155, 164, 165] for a discussion of the relationship between E_b and the potential near the nucleus of the analysed atom and spectra characteristics in NMR, NQR, Mossbauer spectroscopy. The relationship of E_b with basicity in the gaseous phase or relationship with the proton affinities is considered in [166–168]. See also bibliography in [14].

It has been indeed illustrated [71, 72, 169] that the energies O 1s, F 1s, and Cl 2p stand in a roughly linear relationship with $1/R$, where R is the average distance from the neighbour atoms. This is exemplified by data for the O 1s line in M_2O_3 oxides in Fig. 1.12.

The correlation of the values $\Delta E(A)$ and $1/R_{AB}$, as follows from the above analysis, should be applied only to negatively charged A atoms (negative shifts of ΔE_b). This correlation does not seem probable for positive shifts of B atoms because with the growth of A anions, both members, kq_B and $\sum(A)q_A/R_{AB}$ diminish, the kq playing the main role in determing the sign of $E_b(B)$. Hence, the derivative $\partial(\Delta E)/\partial(1/R)$ for positive shifts has a different sign from what it has for the negative (with due account of the difference between the signs of ΔE).

References

1. Jenkin, J. G., Leckey, R. C. G., and Liesegang, J. (1977) The development of X-ray spectroscopy, 1900–1960. *J. Electron Spectrosc. Rel. Phen.* **12**, 1–36.
2. Siegbahn, K., Nordling, C., Fahlman, A. *et al.* (1967) ESCA: atomic molecular and solid state structure studied by means of electron spectroscopy. *Nova Acta Regiae Soc. Sci. Ups.* Ser. IV, **20**, 260 pp.
3. Siegbahn K., Nordling C., Johansson G. *et al.* (1969) *ESCA applied to free molecules.* North Holland, Amsterdam., 232 p.
4. Nefedov, V. I. (1973) *Application of X-ray photoelectron spectroscopy in chemistry.* Molecular structure and chemical bonding series, VINITI, Moscow, (in Russian).
5. Nefedov, V. I. (1975) *Electron levels of chemical compounds.* Molecular structure and chemical bonding series, VINITI, Moscow, (in Russian).
6. Nemoshkalenko, V. V. and Alyoshin, V. G. (1976) *Electron spectroscopy of crystals.* Nauksova dumka, Kiev, 1976. 336 p. In Russian.
7. Brundle, C. R. and Baker, A. D. L., Eds (1977, 78, 80) *Electron spectroscopy: theory, techniques and applications* (5 Vols). Academic Press, New York.
8. Carlson, T. A. (1976) *Photoelectron and auger spectroscopy.* Plenum Press, New York.
9. Feuerbacher, B., Fitton, B., and Willis, R. F., Eds (1978) *Photoemission and electronic properties of surfaces* Wiley, New York.
10. Wagner, C. D., Riggs, W. M. *et al.* (1979) *Handbook of X-ray photoelectron spectroscopy.* Physical Electronics Division, Perkin-Elmer Corpsation, Eden Prairie, MN.
11. Ley, L. and Cardona, M. B., Eds (1979) *Photoemission in solids Topics in Applied Physics*, Vol. 26/27, p. 290. Springer, Berlin.
12. Fadley, C. S. (1978) Basic concepts of X-ray photoelectron spectroscopy. In: *Electron spectroscopy: theory, techniques and applications*, Vol. 2, Academic Press, New York, pp. 9–156.
13. Minachev, Kh. M., Antoshin, G. V., and Shpiro, Y. S. (1981) *Photoelectron spectroscopy and its application in catalysis.* Nauka, Moscow (in Russian).
14. Nefedov, V. I. (1984) *X-ray photoelectron spectroscopy of chemical compounds. Handbook.* Khimiya, Moscow (in Russian).
15. Afanasyev, V. P. and Yavor, S. Y. (1978) *Electrostatic energy analysers for beams of charged particles.* Atomizdat, Moscow (in Russian).
16. Fridrikhov, S. A. (1978) *Energy analysers and monochromators for electron spectroscopy.* Leningrad State University Publishing House (in Russian).

17. Briggs, D. and Seah, M. P., Eds (1983) *Practical surface analysis by Auger and XPS*. Wiley, Heyden.
18. Krause, M. O. and Ferreira, J. G. K. (1975) X-ray emission spectra of Mg and Al. *J. Phys.* **B8**, 2007–2014.
19. Van Attekum, P. M. Th. M. and Trooster, J. M. (1977) Removal of X-ray satellites from Mg K_α excited photoelectron spectra. *J. Electron Spectrosc. Rel. Phen.* **11**, 363–370.
20. Hovland, C. T. (1977) Scanning ESCA – new dimension for electron spectroscopy. *Appl. Phys. Lett.* **30**, 274–275.
21. Yates, K. and West, R. H. (1983) Small area XPS. *Surf. Interface Anal.* **5**, 217–221.
22. Helmer, J. C. and Weichert, N. H. (1968) Enhancement of sensitivity in ESCA spectrometers. **13**, 266–268.
23. Castle, J. E. and West, R. H. (1980) Sensitivity factors, cross-section and resolution data for use with the Si K_α X-ray source. *J. Electron Spectrosc. Rel. Phen.* **19**, 409–428.
24. Madey, T. E., Wagner, C. D., and Joshi, A. (1977) Surface characterization of catalysts using electron spectroscopies: results of a round-robin sponsored by ASTM committee D-32 on catalysts. *J. Electron Spectrosc. Rel. Phen.* **10**, 359–388.
25. Baer, Y. (1976) Natural energy scale for XPS spectra of metals. *Solid State Commun.* **19**, 669–671.
26. Evans, S. (1973) Work functions measurements by X = PE spectroscopy and their relevance to calibration of X = PE spectra. *Chem. Phys. Lett.* **23**, 134–138.
27. Connor, J. A., Considine, M., Hillier, I. H., and Briggs, D. (1977) Low energy photoelectron spectroscopy of solids. Aspects of experimental methodology concerning metals and insulators. *J. Electron. Spectrosc. Rel. Phen.* **12**, 143–150.
28. McCreary, J. R. and Thorn, R. J. (1976) Electrical charging of non-conductors in a photoelectron spectrometer: effects of X-ray source voltage. *J. Electron Spectrosc. Rel. Phen.* **8**, 425–436.
29. Ebel, M. F. and Ebel, H. (1974) Charging effect in X-ray photoelectron spectroscopy. *J. Electron Spectrosc. Rel. Phen.* **3**, 169–180.
30. Freund, H. J., Gonska, H., Lohneiss, H., and Hohlneicher, G. (1977) Charging and rotational dependence of line position. *J. Electron Spectrosc. Rel. Phen.* **12**, 425–434.
31. Gonska, H., Freund, H. J., and Hohlneicher, G. (1977) On the importance of photoconduction in ESCA experiments. *J. Electron Spectrosc. Rel. Phen.* **12**, 435–441.
32. Jaegle, A., Kalt, A., Nanse, G., and Peruchetti, J. C. (1978) Contribution a l'étude de l'effect de charge sur échantillon ésolant en spectroscopie de photoelectrons (XPS). *J. Electron Spectrosc. Rel. Phen.* **13**, 175–186.
33. Wagner, C. D. (1980) Studies of the charging of insulators in ESCA. *J. Elèctron Spectrosc. Rel. Phen.* **18**, 345–349.
34. Brandt, E. S., Untereker, D. F., Reilley, C. N., and Murray, R. W. (1978) A comparison of carbon contaminant build-up on conductors and insulators in X-ray photoelectron spectroscopy. *J. Electron Spectrosc. Rel. Phen.* **14**, 113–120.
35. Nefedov, V. I., Salyn, Ya. V., Leonhardt, G., and Sheibe, R. (1977) A comparison of different spectrometers and charge corrections used in X-ray photoelectron spectroscopy. *J. Electron Spectrosc. Rel. Phen.* **10**, 121–124.
36. Nefedov, V. I., Lenenko, V. S., Shur, V. B. *et al.* (1973) Molecular nitrogen as a ligand X-ray photoelectron spectroscopy study of dinitrogen complexes. *Inorg. Chim. Acta*, **7**, 499–502.
37. Nefedov, V. I. (1982) A comparison of results of an ESCA study of nonconducting solids using spectrometers of different constructions. *J. Electron Spectrosc. Rel. Phen.* **25**, 29–48.

38. Kohiki, S. and Oku, K. (1984) Problems of adventitious carbon as an energy reference. *J. Electron Spectrosc. Rel. Phen.* **33**, 375–380.
39. Nordberg, R., Brecht, U., Albridge, R. G. *et al.* (1970) Binding energy of 2p electrons of silicon in various compounds. *Inorg. Chem.* **9**, 2469–2474.
40. Swift, P. (1982) Adventitious carbon—the panacea for energy referencing? *Surf. Interface Anal.* **4**, 47–51.
41. Hnatowich, D. J., Hudis, J., Perlman, M. L., and Ragaini, R. C. (1971) Determination of charging effect in photoelectron spectroscopy of non-conducting solids. *J. Appl. Phys.* **42**, 4883–4885.
42. Grinnard, C. R. and Riggs, W. M. (1974) X-ray photoelectron spectroscopy of fluoropolymers. *Anal. Chem.* **46**, 1306–1308.
43. Uwamino, Y., Ishizuku, F., and Yamatera, H. (1981) Charge correction by gold deposition onto nonconducting samples in XPS. *J. Electron. Spectrosc. Rel. Phen.* **23**, 55–62.
44. Betteridge, C., Carver, J. C., and Hercules, D. M. (1973) Devaluation of the gold standard in X-ray photoelectron spectroscopy. *J. Electron Spectrosc. Rel. Phen.* **2**, 327–334.
45. Hohiki, S. (1984) Problem of evaporated gold as an energy reference in XPS. *Appl. Surf. Sci.* **17**, 497–503.
46. Mizokawa, Y., Iwasaki, H., Nishitani, R., and Nakamura, S. (1978) ESCA studies of Ga, As, GaAs, Ga_2O_3, As_2O_3, and As_2O_5. *J. Electron Spectrosc. Rel. Phen.* **14**, 129–141.
47. Kohiki, S., Ohmura, T., and Kusao, K. (1983) A new charge correction method in X-ray photoelectron spectroscopy. *J. Electron Spectrosc. Rel. Phen.* **28**, 229–237.
48. Kohiki, S., Ohmura, T., and Kusao, K. (1983) Appraisal of new charge correction method in X-ray photoelectron spectroscopy. *J. Electron Spectrosc. Rel. Phen.* **31**, 85–90.
49. Bremser, W. and Linemann, F. (1971) Problem of calibration in X-ray photoelectron spectroscopy. *Chem. Zg.* **95**, 1011–1013.
50. Halett, L. and Carlson, T. A. (1971) Measurements of chemical shifts in photoelectron spectra of arsenic and bromine compounds. *Appl. Spectrosc.* **25**, 33–36.
51. Stec, W. J., Morgan, W. E., Albridge, R. G., and Van Waser, J. R. (1972) Measured binding energy shifts of 3p and 3d electrons in arsenic compounds. *Inorg. Chem.* **11**, 219–225.
52. Johansson, G., Hedmann, J., Berndtsson, A. *et al.* (1973) Calibration of electron spectra. *J. Electron Spectrosc. Rel. Phen.* **2**, 295–317.
53. Nefedov, V. I. and Kakhana, M. N. (1972) Choice of standard in X-ray photoelectron method of analysis. *Anal. Khim.* **27**, 1249–1252 (in Russian).
54. Huchital, D. A. and McKein, R. T. (1972) Use of an electron flood gun to reduce surface charging in X-ray photoelectron spectroscopy. *Appl. Phys. Lett.* **20**, 158–159.
55. Powell, C. J., Erickson, N. E., and Madey, T. E. (1979) Results of a joint AUGER/ESCA Round Robin sponsored by ASTM committee E = 42 on surface analysis. *J. Electron Spectrosc. Rel. Phen.* **17**, 361–404.
56. Gudat, W. and Eastman, D. E. (1976) Electronic surface properties of 3–5 semiconductor–excitonic effects, band bending effects and interaction with Au and O adsorbate layers. *J. Vac. Sci. Technol.* **13**, 831–837.
57. Sharma, J., Staley, R. M., Rimstidt, J. D. *et al.* (1971) Effect of doping on X-ray photoelectron spectra of semiconductors. *Chem. Phys. Lett.* **9**, 564–567.
58. Hedman, J., Bayer, Y., Berndtsson, A. *et al.* (1972) Influence of doping on the electron spectrum of silicon. *J. Electron Spectrosc.* **1**, 101–104.

59. Ley, L., Pollak, R. A., McFeely, F. R. *et al.* (1974) Total valence band densities of state of III–V and II–VI compounds from X-ray photoemission spectroscopy. *Phys. Rev.* **9**, 600–621.
60. Grandtke, T. and Cardona, M. (1980) Electronic properties of clean and oxygen-covered (100) cleaved surfaces of PbS. *Surf. Sci.* **92**, 385–392.
61. Brillson, L. J., Brucker, C. F., Katnani, A. D. *et al.* (1982) Fermi-level pinning and chemical structure of InP–metal interfaces. *J. Vac. Sci. Technol.* **21**, 564–569.
62. Lewis, R. T. and Kelly, M. A. (1980) Binding energy reference in X-ray spectroscopy of insulators. *J. Electron Spectrosc. Rel. Phen.* **20**, 105–116.
63. Bird, R. J. and Swift, P. (1980) Energy calibration in electron spectroscopy. *J. Electron Spectrosc. Rel. Phen.* **21**, 227–240.
64. Asami, K. (1976) A precisely consistent energy calibration method for X-ray spectroscopy. *J. Electron Spectrosc. Rel. Phen.* **9**, 469–478.
65. Richter, K. and Peplinski, S. (1978) Energy calibration of electron spectroscopy. *J. Electron Spectrosc. Rel. Phen.* **13**, 69–71.
66. Anthony, M. T. and Seah, M. P. (1984) XPS: energy calibration of electron spectrometers. *Surf. Interface Anal.* **6**, 95–115.
67. Nefedov, V. I. (1978) Influence of transition element oxidation state on its X-ray photoelectron spectra. *Koord. Khim.* **4**, 1283–1291 (in Russian).
68. Nefedov, V. I., Salyn, Ya. V. (1978) ESCA study of Pt(III) and Pt(I) compounds. *Inorg. Chim. Acta* **28**, L135–L136.
69. Salyn, Y. V., Nefedov, V. I., Mayorova, A. G., and Kuznetsova, G. M. (1978) X-ray photoelectron study of Pt(III) compounds with acetamide. *Zh. Neorg. Khim.* **23**, 829–831 (in Russian).
70. Barinsky, R. L. and Nefedov, V. I. (1966) *X-ray spectral determination of effective atomic charge.* Nauka, Moscow (in Russian).
71. Nefedov, V. I., Gati, G., Dzhurinsky, B. F. *et al.* (1975) ESCA study of oxides of some elements. *Zh. Neorg. Khim.* **20**, 2307–2314 (in Russian).
72. Nefedov, V. I., Sergushin, N. P., Salyn, Ya. V. *et al.* (1976) X-ray photoelectron study of oxides and molibdates. *J. Microsc. Spectrosc. Electron.* **1**, 551–570.
73. Sergushin, N. N., Shabanova, I. I., Kolobova, K. M. *et al.* (1973) Study of electron structure of iron, cobalt and nickel monosilicides by methods of X-ray photoelectron and X-ray spectroscopy. *Fiz. Met. Metalloved.* **35**, 947–952 (in Russian).
74. Grim, S. O., Matienzo, L. J., and Swartz, E. W. (1977) X-ray photoelectron spectroscopy of some nickel ditiolate complexes. *J. Am. Chem. Soc.* **94**, 5116–5122.
75. McIntyre, N. S. and Cook, M. G. (1975) X-ray photoelectron studies of some oxides and hydrooxides of Co, Ni, Cu. *Anal. Chem.* **47**, 2208–2213.
76. Kim, K. S. and Winograd, N. (1974) X-ray photoelectron spectroscopic studies of nickel–oxygen surfaces using oxygen and argon. *Surf. Sci.* **43**, 625–643.
77. Allen, G. C., Tucker, P. M., and Wild, R. K. (1977) High-resolution Auger electron and X-ray photoelectron spectroscopy in the oxidation of nickel metal. In: Proceedings of the Seventh International Vacuum Congress, Vienna, pp. 959–962.
78. Shirley, D. A., Martin, R. L., Kowalczyk, S. P. *et al.* (1977) Core binding energies of the first thirty elements. *Phys. Rev.* **B15**, 544–552.
79. Povey, A. F. and Sherwood, P. M. (1974) Covalent character of lithium compounds studies by X-ray photoelectron spectroscopy. *J. Chem. Soc. Faraday Trans. II* **70**, 1240–1246.
80. Mavel, G., Escard, J., Costa, P., and Castaing, J. (1973) ESCA surface study of metal borides. *Surf. Sci.* **35**, 109–116.

81. Barrie, A. and Street, F. J. (1975) Auger and X-ray photoelectron spectroscopic study of sodium metal and sodium oxide. *J. Electron. Spectrosc. Rel. Phen.* **7**, 1–31.
82. Wagner, C. D. and Biloen, P. (1973) X-ray excited Auger and photoelectron spectra of partially oxidized magnesium surfaces: the observation of abnormal chemical shifts. *Surf. Sci.* **35**, 82–95.
83. Barrie, A. (1973) X-ray photoelectron spectra of aluminium and oxidized aluminium. *Chem. Phys. Lett.* **19**, 109–113.
84. Petersson, L. G. and Carlsson, S. E. (1977) Clean and oxygen-exposed potassium studied by photoelectron spectroscopy. *Phys. Sci.* **16**, 425–431.
85. Van Doveren, H. and Verhoeven, J. A. Th. (1980) XPS spectra of Ca, Sr, Ba and their oxides. *J. Electron Spectrocs. Rel Phen.* **21**, 265–273.
86. Westerhof. A. and Liefelde Mejer, H. J. (1978) XPS of some organoscandium compounds. *J. Organomet. Chem.* **144**, 61–64.
87. Nefedov, V. I., Salyn, Y. V., Chertkov, A.A., and Padurets, L. N. (1974) X-ray photoelectron study of electron density distribution in transition element hydrides. *Zh. Neorg. Khim.* **19**, 1443–1445 (in Russian).
88. Hamrin, K., Nordling, C., and Kihlborg, L. (1970) ESCA studies of oxidation states in W–V–O systems. Preprint, Uppsala University Institute of Physics, No. 692, March.
89. Bouyssoux, G., Romand, M., Poloschegg, H. D., and Calow J. T. (1977) XPS and AES studies of anodic passive films grown on chromium electrodes in sulphuric acid baths. *J. Electron Spectrosc. Rel. Phen.* **11**, 185–196.
90. Borges–Soares, M., Menes, F., Fontaine, R., and Gaillat, R. (1983) Photoelectron spectroscopy studies of the reactions of Mn with oxygen and water vapour. *J. Microsc. Spectrosc.* **8**, 93–107.
91. Mills Sullivan, J. L. (1983) A study of the core level electrons in iron and its three oxides by means of XPS. *J. Phys.* **D16**, 723–732.
92. Brundle, C. R., Chuang, T. J., and Rice, D. W. (1976) X-ray photoelectron study of the interaction of oxygen and air with clean surface. *Surf. Sci.* **60**, 286–300.
93. Gaarenst, S. W. and Winograd, N. (1977) Initial and final-stage effects in ESCA spectra of cadmium and silver oxides. *J. Chem. Phys.* **67**, 3500–3506.
94. Morgan, W. E. and van Wazer, J. R. (1973) Binding energy shifts in the X-ray photoelectron spectra of a series of related Group IV a compounds. *J. Phys. Chem.* **77**, 964–969.
95. Schwartz, W. E., Kenneth, J., Wynn, K. J., and Hercules, D. M. (1971) X-ray photoelectron spectroscopic investigation of Group VI a elements. *Anal. Chem.* **43**, 1884–1887.
96. Ebbinghaus, G. and Simon, A. (1979) Electronic structure of Rb, Cs and some of their metallic oxides studied by photoelectron spectroscopy. *Chem. Phys.* **43**, 117–133.
97. Cimino, A. and Angelis, B. A. (1980) On the relationship between binding energy and oxidation state of Mo in molybdenium oxides. *J. Catal.* **61**, 182–184.
98. Patterson, T. A., Carver, J. C., Leyden, D. E., and Hercules, D. M. (1976) Surface study of cobalt–molybdena–alumina catalysts using X-ray photoelectron spectroscopy. *J. Phys. Chem.* **80**, 1700–1708.
99. Folkesson, B. (1973) ESCA studies of charge distribution in some dinitrogen complexes of Re, Ir, Ru and Os. *Acta Chem. Scand.* **27**, 287–302.
100. Brinen, J. S. and Mellera, A. J. (1972) ESCA studies on catalysts. Rhodium on charcoal. *J. Phys. Chem.* **76**, 2525–2531.
101. Barr, T. L. (1978) ESCA study of termination of passivation of elemental metals. *J. Phys. Chem.* **82**, 1801–1810.

102. Salyn, Y. V., Starchevsky, M. K., Stolyarov, N. P. *et al.* (1983) X-ray photoelectron spectra of palladium catalysts for oxidizing acetooxylation of olefins and aromatic hydrocarbons. *Kinet. Catal.* **24**, 743–746 (in Russian).
103. Lin, A. W. C., Armstrong, N. R., and Kuwana, T. (1977) X-ray photoelectron Auger electron spectroscopic studies of tin and indium metal foil oxides. *Anal. Chem.* **49**, 1228–1285.
104. Bouderille, Y., Fiqueras, F., Forissier, M. *et al.* (1979) Correlation between X-ray photoelectron spectroscopy data and catalytic properties in selective oxidation on Sb–Sn–O catalysts. *J. Catal.* **58**, 57–60.
105. McGuire, G. E., Schweitzer, D. K., and Carlson, T. A. (1973) Study of core electron binding energies in some Group IIIa, Vb, and VIb compounds. *Inorg. Chem.* **12**, 2450–2453.
106. Haber, J., Stoch, J., and Ungier, L. (1976) Electron spectroscopic studies of the reduction of WO_3. *J. Solid State Chem.* **19**, 113–115.
107. Cimino, A., Angelis, B. A. de, Gazzoli, D., and Valigli, M. (1980) Photoelectron spectroscopy and thermogravimetry of pure and supported rhenium oxides. *Z. Anorg. Allg. Chem.* **460**, 86–98.
108. Kim, K. S., Winograd, N., and Davis, R. E. (1971) Electron spectroscopy of platinum oxygen surfaces and application to electrochemical studies. *J. Am. Chem. Soc.* **93**, 6296–6297.
109. Kim, K. S., O'Leary, T. J., and Winograd, N. (1973) X-ray photoelectron spectra of lead oxides. *Anal. Chem.* **45**, 2214–2218.
110. Morgan, W. E., Stech, W. J., and Wazer, J. R. (1973) Inner-orbital binding energy of Sb and Bi compounds. *Inorg. Chem.* **12**, 953–955.
111. Allen, G. C. and Tucker, P. M. (1973) Surface oxidation of uranium metals as studied by X-ray photoelectron spectroscopy. *J. Chem. Soc. Dalton Trans.* **5**, 470–473.
112. Larson, D. T. and Haschke, J. M. (1981) XPS–AES characterization of plutonium oxides and oxide carbide. *Inorg. Chem.* **20**, 1945–1950.
113. Nefedov, V. I. (1975) Application of X-ray photoelectron spectroscopy for investigation of coordination compounds. *Koord. Khim.* **1**, 291–318 (in Russian).
114. Nefedov, V. I., Salyn, Y. V., and Kohler, H. (1979) X-ray photoelectron studies of lead and mercury compounds. *Zh. Neorgan. Khim.* **24**, 2564–2566 (in Russian).
115. Nefedov, V. I. and Porai-Koshits, M. A. (1972) Additive nature of inner levels shifts of atomic in chemical compounds. *Zh. Strukt. Khim.* **13**, 865–868 (in Russian).
116. Gelius, U., Heden, P. T., Hedman, J. *et al.* (1970) Molecular spectroscopy by means of ESCA. III. Carbon compounds. *Phys. Scr.* **2**, 70–80.
117. Lindberg, B. J. and Hedman, J. (1975) Molecular spectroscopy by means of ESCA. VI. Group shifts for N, P and As compounds. *Chem. Scr.* **7**, 155–166.
118. Gray, R. C., Carver, J. C., and Hercules, D. M. (1976) An ESCA study of organosilicon compounds. *J. Electron Spectrosc. Rel. Phen.* **8**, 343–358.
119. Fluck, E. and Weber, D. (1975) Application of X-ray photoelectron spectroscopy in phosphorus chemistry. *Z. Anorg. Allg. Chem.* **412**, 47–58.
120. Feltham, R. D. and Brant P. (1982) XPS studies of core binding energies in transition metal complexes. *J. Am. Chem. Soc.* **104**, 641–645.
121. Jolly, W. L. and Bakke, A. A. (1976) Prediction of core electron-binding energies with a four-parameter equation. *J. Am. Chem. Soc.* **98**, 6500–6504.
122. Lindberg, B. J., Hamrin, K., Johansson, G. *et al.* (1976) Molecular spectroscopy by means of ESCA. II. Sulfur compounds. *Phys. Scr.* **1**, 286–298.
123. Salyn, Y. V., Titova, K. V., Loginova, Ye. N. *et al.* (1977) X-ray photoelectron study of nitrogen compounds. *Zh. Neorg. Khim.* **22**, 2998–3003 (in Russian).

124. Nefedov, V. I., Kokunov, Y. V., and Buslayev, Y. A. (1974) X-ray photoelectron study of alkaline and alkaline earth metal fluorides. *Zh. Neorg. Khim.* **19**, 1166–1169 (in Russian).
125. Nefedov, V. I. (1977) X-ray photoelectron spectra of halogens in coordination compounds. *J. Electron Spectrosc. Rel. Phen.* **12**, 459–476.
126. Haber, J., Stock, J., and Ungier, L. (1976) X-ray photoelectron spectra of oxygen in oxides of Co, Ni, Fe and Zn. *J. Electron Spectrosc. Rel. Phen.* **9**, 459–468.
127. Au, C. T. and Roberts, H. W. (1980) Photoelectron spectroscopic evidence for the activation of adsorbate bonds by chemisorbed oxygen. *Chem. Phys. Lett.* **74**, 472–474.
128. Alyoshin, V. G. and Kharlamov, A. I. (1980) X-ray photoelectron spectra of zirconium, niobium and molybdenum sulphides. *Zh. Neorgan. Khim.* **25**, 2036–2039 (in Russian).
129. Mavel, G., Escard, J., Costa, P., and Castaing, J. (1973) ESCA surface study of metal borides. *Surf. Sci.* **35**, 109–116.
130. Wagner, C. D., Gale, L. H., and Ragmond, R. H. (1979) Two-dimensional chemical state plots: a standardized data set for use in identifying chemical states by X-ray photoelectron spectroscopy. *Anal. Chem.* **51**, 466–482.
131. Hercules, D. M. (1972) Electron spectroscopy. II. X-ray excitation. *Anal. Chem.* **44**, 106R–112R.
132. Hercules, D. M. and Carver, J. C. (1974) Electron spectroscopy: X-ray and electron excitation. *Anal. Chem.* **46**, 133R–150R.
133. Hercules, D. M. (1976) Electron spectroscopy! X-ray and electron excitation. *Anal. Chem.* **48**, 294R–313R.
134. Baker, A. D., Brisk, M. A., and Liotta, D. C. (1978) Electron spectroscopy: ultraviolet and X-ray excitation. *Anal. Chem.* **50**, R328–R346.
135. Baker, A. D., Brisk, M. A., and Liotta, D. C. (1980) Electron spectroscopy: ultraviolet and X-ray excitation. *Anal. Chem.* **52**, R161–R174.
136. Bowling, R. A. and Larrabee, G. B. (1983) Surface characterization. *Anal. Chem.* **55**, 133R–156R.
137. Turner, N. H., Dunlop, B. L., and Colton, N. J. (1984) Surface analysis review. *Anal. Chem.* **56**, 373R–416R.
138. Nefedov, V. I. (1967) Electron structure of molecule on the basis of X-ray spectral data. *Zh. Strukt. Khim.* **8**, 1037–1042 (in Russian).
139. Fadley, C. S., Hagstrom, S. B., Klein, M. P., and Shirley, D. A. (1968) Effects on core-electron binding energies in iodine and europium. *J. Chem. Phys.* **48**, 3779–3782.
140. Gelius, U., Roos, B., and Siegbahn, P. (1970) MO SCF-LCAO studies of sulfur compounds. *Chem. Phys. Lett.* **4**, 471–475.
141. Nordberg, R., Gelius, U., Heden, P. F. *et al.* (1968) Charge distribution in carbon compounds. Preprint UUIP-581, Uppsala University Institute of Physics, p. 37.
142. Politzer, P. and Politzer, A. (1973) An easy procedure for estimating atomic charges from calculated core-electron energies. *J. Am. Chem. Soc.* **95**, 5450–5455.
143. Jolly, W. L. and Perry, W. P. (1973) Estimation of atomic charges by an electronegativity equalization procedure calibrated with core binding energies. *J. Am. Chem. Soc.* **95**, 5442–5450.
144. Parry, D. E. (1975) Determination of atomic partial charges using X-ray photoelectron spectroscopy—application to crystalline solids. *J. Chem. Soc. Faraday Trans. II* **72**, 337–345.
145. Fellner-Felldegg, H. (1976) Elektronenspektroskopie heute und morgen. *Wiss. Z. Karl-Marx-Univ. Leipzig* **4**, 355–382.

146. Pireaux, J. J., Swensson, S., Basilier, E. *et al.* (1976) Valence band formation and electronic relaxation. I. X-ray photoelectron spectra of linear alkanes. *Phys. Rev.* **A14**, 2133–2145.
147. Martin, R. L. and Shirley, D. A. (1974) Relation of core level binding energy shifts to proton affinity and Lewis basicity. *J. Am. Chem. Soc.* **96**, 5299–5304.
148. Ley, L., Kowalczyk, S. P., McFeely, F. R. *et al.* (1973) X-ray photoemission from zinc—evidence for extraatomic relaxation via semilocalized excitons. *Phys. Rev.* **B8**, 2392–2408.
149. Watson, R. E., Perlman, M. L., and Herbst, J. F. (1976) Core level shifts in 3d transition metals and tin. *Phys. Rev.* **B13**, 2358–2365.
150. Broughton, J. Q. and Perry, D. L. (1979) An equivalent cores calculation of solid-state and adsorbate shifts in X-ray photoelectron spectroscopy. *J. Electron Spectrosc. Rel. Phen.* **16**, 45–64.
151. Hermann, K. and Bagus, P. S. (1977) Binding and energy-level shifts of carbon monoxide adsorbed on nickel: model studies. *Phys. Rev.* **16**, 4195–4208.
152. Goscinski, O., Hehenberger, M., Roos, B., and Siegbahn, P. (1975) Transition operators for molecular ΔE SCF calculations. Ionization in water and furan. *Chem. Phys. Lett.* **33**, 427–431.
153. Howat, G. and Goscinski, O. (1975) Relaxation effects on ESCA chemical shifts by a transition potential model. *Chem. Phys. Lett.* **30**, 87–90.
154. Schwartz, M. E. (1970) Correlation of ls binding energy with the average quantum mechanical potential at a nucleus. *Chem. Phys. Lett.* **6**, 631–636.
155. Schwartz, M. E. (1970) Correlation of core electron binding energies with the average potential at a nucleus. *Chem. Phys. Lett.* **7**, 78–82.
156. Aitken, E. J., Bahl, M. K., Bomben, K. D. *et al.* (1980) Electron spectroscopic investigation of the influence of initial and final state effects on electronegativity. *J. Am. Chem. Soc.* **102**, 4873–4879.
157. Bechstedt, F., Enderlein, R., Fellenberg, R. *et al.* (1983) A new method for determining relaxation energies by means of AES and XPS and its application to silicon compounds. *J. Electron Spectrosc. Rel. Phen.* **31**, 131–144.
158. Jolly, W. L. and Hendrickson, D. (1970) Thermodynamic interpretation of chemical shifts in core electron binding energies. *J. Am. Chem. Soc.* **92**, 1863–1866.
159. Jolly, W. L. (1970) A new method for estimation of dissociation energies and its application to correlation of core electron binding energies obtained from X-ray photoelectron spectra. *J. Am. Chem. Soc.* **92**, 3260–3263.
160. Bagus, P. S. and Schaefer, H. F. (1972) Localized and delocalized ls hole states of O_2^+ molecular ions. *J. Am. Phys.* **56**, 224–228.
161. Jolly, W. L. (1977) Inorganic applications of X-ray photoelectron spectroscopy *Top. Curr. Chem.* **71**, p. 150–182.
162. Shirley, D. A. (1972) Near equivalence of quantum mechanical potential model and thermochemical model of ESCA shifts. *Chem. Phys. Lett.* **15**, 325–328.
163. Tomanek, D., Dowden, P. A., and Grunze, M. (1983) Thermodynamic interpretation of core-level binding energies in adsorbates. *Surf. Sci.* **126**, 112–119.
164. Gelius, U. (1974) Binding energies and chemical shifts in ESCA. *Phys. Scr.* **9**, 133–147.
165. Shirley, D. A. (1975) Hyperline interactions and ESCA data. *Phys. Scr.* **11**, 117–120.
166. Mills, B. E., Martin, R. L., and Shirley, D. A. (1976) Further studies of core binding energy–proton affinity correlation in molecules. *J. Am. Chem. Soc.* **98**, 2380–2385.
167. Davis, D. W. and Rabalais, J. W. (1974) Model for proton affinities and inner shell electron binding energies based on the Hellmann–Feynman theorem. *J. Am. Chem. Soc.* **96**, 5305–5310.

168. Brown, R. S. and Tse, A. (1980) Determination of circumstances under which the correlation of core binding energy and gas phase basicity or proton affinity breakdowns. *J. Am. Chem. Soc.* **102**, 5222–5226.
169. Nefedov, V. I., Sergushin, N. P., and Salyn, Ya. V. (1976) Correlation between chemical shifts of X-ray photoelectron spectra and interatomic distances. *J. Electron Spectrosc. Rel. Phen.* **8**, 81–84.
170. Wandelt, K. (1982) Photoemission studies of adsorbed oxygen and oxide layers. *Surf. Sci. Rep.* **2**, 1–121.

Chapter 2

Line intensity and quantitative analysis

2.1. Dependence of intensity of X-ray photoelectron lines on various factors

The intensity I (the number of electrons reaching the detector per unit of time) of X-ray photoelectron lines can be used in making a quantitative analysis of the surface of solids. Let us consider the dependence of I on various factors. This question has been studied in detail in [1–8].

The intensity I of X-ray photoelectron lines can be presented as a product of three factors: instrumentation A, process of photoionization B, and characteristics of sample C:

$$I = ABC \tag{2.1}$$

2.1.1. *Instrumentation factor*

The intensity of X-ray radiation in layer dz at depth z (see Fig. 1.4) is determined by the equation

$$I_z = I_0 \exp(-z/\lambda_x \sin\Phi') \approx I_0, \tag{2.2}$$

where λ_x is the linear absorption coefficient of the X-rays. Since $\lambda_x \gg \lambda_e$, where λ_e is the mean free path in a solid and the depth z of the sample from which photoelectrons can still reach the surface is of the order λ_e then the value z/λ_x is small and the second equality in (2.2) is true. It should be noted that the radiation density per unit of sample surface, I_0, being fixed, is proportional to $\sin\Phi$, but the path of the X-ray radiation in the layer dz is inversely proportional to $\sin\Phi$, hence the probability of photoionization in layer dz does not depend on angle Φ. The only exception is the range of ultimately small values $\Phi \leqslant 1°$, at which the effects of refraction and reflection can increase the relative intensity of the X-ray flux in the topmost surface films [2, 5].

The effective area of the sample S is determined by several factors: the angle α, the area of the entrance slot of the analyser A_0, and the area of the X-ray spot S_0. If the area of the X-ray spot is not a limiting factor, then

$$S = A_0/\sin\alpha \tag{2.3}$$

at $\alpha > \alpha_{min}$, whereas at $\alpha < \alpha_{min}$

$$S = A_0/\sin\alpha_{min}, \tag{2.4}$$

where α_{min} is the smallest angle at which the spectrometer's slit is still fully 'closed' by the sample. If the area of the X-ray spot is the limiting factor, then

$$S = S_0/\sin \Phi. \tag{2.5}$$

The luminosity of the spectrometer T (see (1.3), (1.7), and (1.8)) is proportional to the body angle Ω_0 and depending on the nature of the analyser is proportional or inversely proportional to E_{kin}:

$$T \sim \Omega_0 E_{kin} \qquad T \sim \Omega_0/E_{kin}. \tag{2.6}$$

So usually we have

$$A = I_0(A_0/\sin \alpha)[\Omega_0(E_{kin})F(E_{kin})], \tag{2.7}$$

where F is the efficiency of the electron detector.

It shall be noted that the dependence of A on α is deduced with some simplifying assumptions*; in particular the dependence of I_0 and Ω_0 on α is ignored. A more detailed analysis [10] shows that even in this case the ratio of intensities of the two lines does not depend in practice on the function $R(\alpha)$, which is determined as follows:

$$R(\alpha) \equiv \sin \alpha \iint I(\alpha, x, y)\Omega(\alpha, x, y)\mathrm{d}x\mathrm{d}y \tag{2.8}$$

(integration is with respect to the area of the sample).

2.1.2. *Photoionization cross-section*

Here we shall dwell on some main results of calculating photoionization cross-sections of interest to X-ray photoelectron spectroscopy (see review in [11]). It should be recalled that the photoionization cross-section of atomic level σ_{nl} has dimensions of cm^2 and yields the ionization probability of the shell nl in atom A when one photon falls on a unit of surface per unit of time. If the intensity of the flux of photons is I and m is the number of A atoms per unit of area, then the number N of photoelectrons from the Anl shell emitted per unit of irradiation surface equals

$$N = Im\sigma_{nl} \tag{2.9}$$

The differential photoionization cross-section $\mathrm{d}\sigma/\mathrm{d}\Omega$, characterizing the flux of photoelectrons per unit of time in the body angle Ω as a result of transition of the N-electron system $\psi^i(N)$ from the initial state to the final one under the impact of photon hν, can be presented by the ratio

$$\mathrm{d}\sigma/\mathrm{d}\Omega = (c/\mathrm{h}\nu)\sum_{i,f} |\bar{\mathbf{e}}\langle \psi^f(N_k)|\sum(k)\bar{\nabla}_k|\psi^i(N_k)\rangle|^2, \tag{2.10}$$

where c is a combination of fundamental constants; $\bar{\mathbf{e}}$ is the unit vector parallel to the electric field vector $\bar{\mathbf{E}}$; and k is the electron index. The matrix element of the

*Use was made of equality (2.3) based on the assumption of the uniform X-ray irradiation of the area of the entire sample at any angles α. If this condition is not met, the dependence A on α may have a complex nature [9, 10].

probability of transition $i \to f$ in (2.10) can be presented in the following three forms:

$$i/\bar{h}\langle\psi^f(N)|\sum(k)\bar{p}_k|\psi^i(N)\rangle = (mhv|\bar{h}^2)\langle\psi^f(N)/\sum(k)\bar{r}_k|\psi^i(N)\rangle$$
$$= 1/hv\langle\psi^f(N)|\sum(k)\bar{\nabla}_k V|\psi^i(N)\rangle, \qquad (2.11)$$

where $V(\bar{r}_k)$ is the potential of electron–electron and electron–nuclear interactions.

These forms of notation are equivalent if precise wave functions are used. The first form is called the equation of 'velocity', the second the equation of 'length' and the third the equation of 'acceleration' in accordance with the operators of matrix elements. The use of the various forms is analysed in [12].

The matrix elements (2.10) and (2.11) are calculated in various approximations. The simplest and most frequently used approximation assumes that photoemission affects only one electron [for clarity designated (1)]; to be more correct, only a single one-electron function changes, this corresponding to a transition from the initial bound state $\varphi^i(1)$ to the state of the continuous spectrum $\varphi^f(1)$. In accordance with this $\psi^i(N)$ and $\psi^f(N)$ equal:

$$\psi^i(N) = |\varphi^i(1)\psi^i(N-1)| \qquad \psi^f(N) = |\varphi^f(1)\psi^f(N-1)|. \qquad (2.12)$$

The notation of functions ψ^i and ψ^f in the form of determinants is assumed in (2.12).

Assuming that the functions of $(N-1)$-electrons do not change for the initial and final states (are 'frozen'), then for the matrix elements we have

$$\langle\psi^f(N)|\sum(k)\bar{t}_k|\psi^i(N)\rangle = \langle\varphi^f(1)|\bar{t}_k|\varphi^i(1)\rangle. \qquad (2.13)$$

A more precise form of notation taking into account the difference of functions* $\psi^i(N-1)$ and $\psi^f(N-1)$, has the form

$$\langle\psi^f(N)|\sum(k)\bar{t}_k|\psi^i(N)\rangle = \langle\varphi^f(1)|\bar{t}_1|\varphi^i(1)\rangle\langle\psi^f(N-1)|\varphi^i(N-1)\rangle. \quad (2.14)$$

The overlap integral in (2.14) is the determinant of overlap integrals of $(N-1)$-electron wave functions. In reference [13] it is shown that the expression (2.14) can be presented with a high degree of precision in the form

$$\langle\varphi^f(N)|\sum(k)\bar{t}_k|\psi^i(N)\rangle = \langle\varphi^f(1)\bar{t}_1|\varphi^i(1)\rangle\Pi(k, k \neq 1)|\langle\varphi^i_k|\varphi^f_k\rangle, \qquad (2.15)$$

where $\Pi(k, k \neq 1)\langle\varphi^i_k|\varphi^f_k\rangle$ is the product of the diagonal elements of the overlap determinant $\langle\psi^f(N-1)|\psi^i(N-1)\rangle$. The expressions (2.14) and (2.15) describe the probability of photoionization in the 'sudden excitation' approximation. They show that several final states $\psi^f(N-1)$ are possible in the event of photoionization $\varphi^i(1) \to \varphi^f(1)$. The necessary condition of the manifestation of the final state is reduced to the difference from zero of the overlap determinant (2.14) or the product of overlap integrals (2.15). This condition can be met not only when $\varphi^i(k)$ equals $\varphi^f(k)$ for any k electron, but also when monopoly selection rules are

*In this case $\psi^i(N-1)$ represents the ground state (in the sudden excitation approximation) while $\psi^f(N-1)$ represents the state with one ejected electron.

exercised in the case of one or several electrons $(N-1)$. So along with the main photoelectron maximum, which accords with the equations $\varphi^i(1)\rightarrow\varphi^f(1)$ and $\psi^i(k)\approx\psi^f(k)$, $k\neq 1$ there is a set of additional monopoly excitation satellites, the intensity of these maxima being proportional to $\langle\psi^f(N-1), m|\psi^i(N-1)\rangle^2$ [see (2.10) and (2.14), where m designates various final states $\psi^f(N-1)$].

Observance of the following important conservation rule is shown in [14].

$$\begin{aligned} &I_{tot}=\sum(m)I_m,\ I_{tot}=c|\langle\varphi^f(1)|\bar{t}_1|\varphi^i(1)\rangle|^2,\\ &I_m=c\sum(m)|\langle\varphi^f(1)|\bar{t}_1|\varphi_1^i(1)\rangle|^2|\langle\psi^f(N-1), m|\psi^i(N-1)\rangle|^2. \end{aligned} \quad (2.16)$$

According to this equation that combined intensity of the X-ray photoelectron transition $\varphi^i(1)\rightarrow\varphi^f(1)$, when the functions of other electrons are 'frozen', is distributed throughout all the monopoly excitation satellites and the principal maximum.

The more general mechanism of the appearance of satellites is connected with the configurations interaction [15]. A lessening of the intensity of the principal maximum at the expense of the intensity of the satellites is also observed in this case.

The well known dipole selection rules are applied to the transition $\varphi^i(1)\rightarrow\varphi^f(1)$:

$$\Delta l=\pm 1,\ \Delta m_l=0,\ \pm 1, \cdot\Delta m_s=0. \quad (2.17)$$

In the case of non-polarized emission the total photoionization cross-section of the atomic shell nl has the form (see (2.10) and (2.13), $\bar{t}=\bar{r}$):

$$\sigma_{nl}=[(4\pi\alpha a_0^2/3)(\mathrm{h}\nu)][lR_{l-1}^2+(l+1)R_{l+1}^2], \quad (2.18)$$

where a is the Bohr radius; α is the fine structure constant; $R_{l\pm 1}=\int R_{nl}(r)\cdot R_{l\pm 1}(r)r^2\mathrm{d}r$; $R_{nl}(r)$ is the radial part of function $\varphi^i(nl)$; and $R_{l\pm 1}(r)$ is the wave function of the photoelectron φ^f with energy $E_{kin}=\mathrm{h}\nu-E_f(nl)$. Usually $R_{l+1}^2\gg R_{l-1}^2$.

The differential cross-section equals

$$\frac{\mathrm{d}\sigma_{nl}}{\mathrm{d}\Omega}=\frac{\sigma_{nl}}{4\pi}[1+1/2\beta(3/2\sin^2\theta-1)], \quad (2.19)$$

where θ is the angle between the directions of the photons $\mathrm{h}\nu$ and the photoelectrons; and β is the asymmetry parameter, which is a function of $R_{l\pm 1}$ and the phases δ_{l+1} characterizing the oscillating nature of $R_{l\pm 1}(r)$. β has a value of 2 for $l=0$ and $-1\leqslant\beta<2$ for other values of l. When $\beta=2$ the value σ_{nl} has its maximum at $\theta=90°$ and equals zero at $\theta=0$ and 180°. When $\beta=-1$ the value σ_{nl} has its minimum at $\theta=90°$ and its maximum at $\theta=0$ and 180°.

The following extensive calculations of σ_{nl} have been published to date for lines Al K_α and Mg K_α for elements $Z\leqslant 20$ in reference [16], for $Z\leqslant 54$ in [17], for $Z\leqslant 92$ in [18], for lines Al K_α, Mg K_α, Y M_ζ, Zr M_ζ, B K_α, C K_α, N K_α, O K_α, F K_α, Co L_α, Ni L_α, Cu L_α, Na K_α, Ti K_α for $Z\leqslant 100$ in [19], for the lines Y M_ζ and Zr M_ζ for $Z\leqslant 36$ in [20, 21], for the lines F K_α at $Z\leqslant 36$ in [21, 22], and for the lines Au M_α and Ag L_α at $Z\leqslant 36$ in [23].

Table (2.1) contains relative values of σ_{nl} and intensities (K 2p line taken for unit) for the most intensive lines for $Z \leqslant 92$.

References [27–30] contain relative line intensities for various types of spectrometers.

The extensive calculations of the value β are published in [31] (Table 2.2). In

Table 2.1
Relative photoionization cross-sections and intensity of X-ray photoelectron lines (values for K 2p Taken for unit)

Level, element	Theory	Experiment		Theory		Experiment	
	[19]	[24]	[25]	[26]	[19]	[24]	[26]
		Mg Kα				Al Kα	
1s							
3Li	0.0147	0.0167	0.0257	—	0.0143	0.0148	—
4Be	0.0493	0.0257	—	—	0.0470	0.0333	—
5B	0.122	0.100	0.120	0.102	0.122	0.0955	0.106
6C	0.247	0.157	0.125	—	0.252	0.192	0.215
7N	0.440	0.300	0.254	0.366	0.456	0.348	0.368
8O	0.710	0.463	0.276	0.563	0.742	0.523	0.599
9P	1.06	0.657	0.328	—	1.13	0.803	—
2s							
11Na	0.0971	0.103	0.145	0.115	0.107	0.117	0.117
12Mg	0.131	0.123	0.181	—	0.146	0.142	0.142
13Al	0.170	—	0.189	0.0983	0.191	0.159	0.141
14Si	0.213	—	0.210	0.117	0.243	0.182	0.167
2p							
11Na	0.0525	—	0.0828	—	0.0489	0.0614	—
12Mg	0.0892	—	0.119	—	0.0842	0.0909	—
13Al	0.141	0.15	0.134	—	0.135	0.114	—
14Si	0.214	0.25	0.263	—	0.205	0.177	—
15P	0.309	0.33	—	—	0.300	0.264	0.256
16S	0.431	0.41	0.472	0.442	0.421	0.418	0.375
17Cl	0.586	0.53	0.581	0.513	0.579	0.582	0.550
19K	1.00	1.00	1.00	1.00	1.00	1.00	1.00
20Ca	1.26	1.13	0.950	0.884	1.27	1.18	1.02
21Sc	1.58	1.20	—	—	1.60	—	1.28
22Ti	1.94	1.15	0.888	—	1.99	1.09	—
23V	2.36	1.33	0.723	—	2.42	1.45	—
24Cr	2.86	1.40	0.894	—	2.95	1.68	—
25Mn	3.35	1.55	—	—	3.40	2.32	—
26Fe	3.93	1.81	0.782	—	4.12	2.77	—
27Co	4.55	2.37	—	—	4.81	3.18	—
28Ni	5.22	2.30	0.772	—	5.57	3.77	—
3p							
24Cr	0.277	—	0.234	0.215	0.296	—	0.172
25Mn	0.331	—	0.256	0.141	0.358	—	0.161
26Fe	0.387	—	0.315	0.200	0.419	—	0.222
27Co	0.445	—	0.446	0.198	0.485	—	0.256
28Ni	0.568	—	0.521	0.278	0.558	—	0.345
29Cu	0.568	0.28	0.472	—	0.629	—	—
30Zn	0.642	0.46	0.591	—	0.716	—	0.488
31Ga	0.723	0.92	—	—	0.813	—	—
32Ge	0.811	—	—	—	0.917	—	—
33As	0.903	—	—	—	1.03	—	—
34Se	1.00	—	—	—	1.15	—	0.882

(*Table contd. over*)

Table 2.1 (*Contd.*)

Level, element	Theory	Experiment		Theory		Experiment	
	[19]	[24]	[25]	[26]	[19]	[24]	[26]
35Br	1.10	—	0.894	—	1.28	—	0.876
3d							
33As	0.485	0.550	—	—	0.456	—	0.345
34Se	0.608	0.567	—	0.878	0.577	0.500	0.571
35Br	0.751	0.700	0.865	0.639	0.716	0.606	—
37Rb	1.10	—	1.33	—	1.06	—	—
38Sr	1.30	1.33	—	—	1.27	—	—
39Y	1.54	—	—	—	1.50	—	—
40Zr	1.79	1.62	—	2.05	1.76	1.36	1.79
41Nb	2.09	1.75	—	1.92	2.07	1.67	1.74
42Mo	2.41	2.10	—	—	2.39	2.27	1.71
43Te	2.73	—	—	—	2.74	—	—
44Ru	3.12	1.75	—	—	3.14	—	—
45Rh	3.53	2.67	—	—	3.57	2.88	—
46Pd	4.00	3.08	—	—	4.05	3.33	—
47Ag	4.45	—	3.41	—	4.55	—	—
48Cd	4.96	5.11	3.82	—	5.08	4.70	—
49In	5.50	5.43	—	—	5.66	5.15	—
50Sn	6.09	5.67	3.33	—	6.31	5.91	—
51Sb	6.73	6.33	—	—	6.98	6.29	—
52Te	7.38	—	—	—	7.70	—	—
53I	8.09	6.67	3.06	—	8.68	6.82	—
54Xe	8.83	—	—	—	19.29	—	—
55Cs	9.54	—	—	—	10.11	—	—
56Ba	10.29	7.83	—	—	10.98	9.33	—
4d							
50Sn	0.633	—	0.960	—	0.677	—	—
51Sb	0.735	—	—	—	0.792	—	1.04
52Te	0.843	—	—	—	0.913	—	0.815
53I	0.958	—	1.11	0.885	1.04	—	—
54Xe	1.08	—	—	—	1/18	—	—
55Cs	1.20	—	1.86	—	1.32	—	1.67
56Ba	1.33	—	2.20	—	1.47	—	1.64
4f							
73Ta	2.23	—	—	2.17	2.16	—	1.96
74W	2.52	1.95	2.93	2.18	2.46	—	1.91
75Re	2.83	2.43	—	2.53	2.77	—	—
76Os	3.15	2.43	—	—	3.11	—	—
77Ir	3.50	—	—	—	3.48	—	4.58
78Pt	3.88	3.13	—	3.31	3.88	—	2.90
79Au	4.30	—	4.72	—	4.30	—	—
80Hg	4.72	4.53	4.43	—	4.73	—	—
81Tl	5.18	4.67	5.57	—	5.23	—	—
82Pb	5.65	5.37	5.36	4.93	5.73	—	5.29
83Bi	6.15	—	—	$4d_{5/2}$ 1.72	6.27	—	5.47
84Po	6.66	—	—	—	6.63	—	—
85At	7.24	—	—	—	7.42	—	—
86Rh	7.81	—	—	—	8.05	—	—
87Fr	8.37	—	—	—	8.68	—	—
88Ra	8.97	—	—	—	9.33	—	—
89At	9.56	—	—	—	10.00	—	—
90Th	10.17	—	8.43	—	10.69	—	—
91Pa	10.92	—	—	—	11.52	—	—
92U	11.61	—	5.41		12.30	—	—

Table 2.2
Values of β

Level	Z	β^a	β^b	Line	Z	β^a	β^b	Line	Z	β^a	β^b
2p	5	0.19	0.38	4p	35	1.62	1.59	4f	60	1.05	1.02
	10	0.77	0.69		40	1.65	1.63		65	1.06	1.04
	15	1.17	1.09		45	1.68	1.67		70	1.05	1.05
	20	1.39	1.35		50	1.69	1.69		75	1.03	1.04
	25	1.46	1.46		55	1.70	1.70		80	1.00	1.03
	30	1.26	1.40		60	1.69	1.70		85	0.95	1.00
					65	1.67	1.69		90	0.88	0.95
3p	15	1.13	1.05		70	1.65	1.68		95	0.75	0.89
	20	1.35	1.28		75	1.61	1.65		100	0.66	0.81
	25	1.49	1.43		80	1.55	1.62				
	30	1.56	1.52		85	1.46	1.56	5p	65	1.73	1.73
	35	1.60	1.58		90	1.28	1.47		70	1.72	1.73
	40	1.61	1.61		95	0.93	1.33		75	1.72	1.73
	45	1.59	1.61						80	1.71	1.72
	50	1.52	1.58	4d	40	1.22	1.16		85	1.70	1.72
	55	1.30	1.44		45	1.27	1.23		90	1.69	1.71
	60	—	1.19		50	1.31	1.28		95	1.67	1.70
					55	1.33	1.32		100	1.65	1.69
3d	25	0.90	0.83		60	1.33	1.33				
	30	1.04	0.98		65	1.31	1.33	5d	75	1.34	1.35
	35	1.14	1.10		70	1.27	1.31		80	1.33	1.35
	40	1.19	1.16		75	1.24	1.27		85	1.31	1.35
	45	1.21	1.20		80	1.15	1.23		90	1.29	1.34
	50	1.17	1.20		85	1.04	1.16		95	1.26	1.32
	55	1.04	1.14		90	0.87	1.06		100	1.23	1.31
	60	0.72	0.97		95	0.63	0.91				
	65	—	0.62		100	0.24	0.69	6p	85	1.72	1.74
									90	1.72	1.74
									95	1.72	1.74
									100	1.72	1.73

[a]For line Mg K_α.
[b]For line Al K_α.

the references quoted above [16–23, 31] the functions $\varphi^i(nl)$ and φ^f were calculated by the Hartree–Fock–Slater relativistic method or by the Hartree–Fock method. The use of a plane wave or an orthogonalized plane wave as φ^f results in a distortion both of the value σ_{nl} and of the dependence $d\sigma_{nl}/d\Omega$ on θ [20, 32]. In a number of instances (usually at small values of E_{kin}) the value σ_{nl} depends substantially on the configuration interaction [33].

So value B in the equation (2.1) is

$$B = K\sigma_{nl}/4\pi[1 + 1/2\beta(3/2\sin^2\theta - 1)], \tag{2.20}$$

where K [see (2.14) and (2.15)] takes into account intensity losses of the main maximum of the A nl line at the expense of satellites. Usually $K = 0.8 \div 0.9$, but there are also instances when $K \sim 0.5$.

2.1.3. *Influence of properties of the sample on intensity*

Let us consider N photoelectrons moving to the surface of the sample. We will take into account only inelastic interactions (for account of elastic interactions see Section 2.2 of this chapter). Let dN be the number of photoelectrons experiencing inelastic interaction when traversing the path dl in a solid. In the case of a homogeneous sample the value dN should be proportional to Ndl and inversely proportional to the value λ of the mean free path of the photoelectron in the given solid without inelastic collisions:

$$dN = -Ndl/\lambda = -Ndz/\lambda \sin\alpha. \tag{2.21}$$

The minus sign is included to account for the withdrawal of dN electrons from the total number of N electrons having an initial energy E_{kin} that accords with the ionization of the nl shell of atom A. As a result of integration (2.21) we have

$$N_z = N_0 \exp(-z/\lambda \sin\alpha), \tag{2.22}$$

where N_z/N_0 is the share of photoelectrons that emerged from the sample (see Fig. 1.4) after passing the path $l = z/\sin\alpha$.

Since the weakening of the X-rays when passing a path of the order of λ is infinitesimally small, it can be held that the entire field of the sample from which the photoelectrons escape is under a uniform impact of X-ray irradiation of I_0 intensity (see above). It can be held, therefore, that in the volume Sdz at a distance z from the surface, at any z of the order of λ, as a result of photoionization of σ_{nl} shell we have dN_0 photoelectrons with a kinetic energy E_{kin} emitted into the body angle Ω_0 per unit of time:

$$dN_0 = \mathscr{T}(d\sigma_{nl}/d\Omega)\Omega_0' nSdl, \tag{2.23}$$

where n is the number of A atoms per unit of volume, and $\mathscr{T}$ is the number of photons falling each second per unit of area.

In accordance with (2.22), of the total number of dN_0 photoelectrons only dN_z electrons reach the surface of the sample without inelastic collisions:

$$dN_z = D\exp(-z/\lambda \sin\alpha)dz, \tag{2.24}$$

where $D = \mathscr{T}(d\sigma_{nl}/d\Omega)\Omega_0 nS$.

Let us consider the following cases:

1. A semi-infinite sample. The number of electrons N that have escaped from the sample can be obtained by integrating (2.24) with respect to z from zero to infinity:

$$N = D\lambda \sin\alpha. \tag{2.25}$$

2. A layer of thickness t. Integration (2.24) is with respect of $z/\sin\alpha$ from zero to $t/\sin\alpha$:

$$N = D[1 - \exp(-t/\lambda \sin\alpha)]\sin\alpha. \tag{2.26}$$

3. A semi-infinite sample beneath a continuous overlayer of thickness t. On the basis of (2.25) and (2.22) for photoelectrons from the sample we have

$$N = D\lambda_1 \exp(-t/\lambda_2 \sin\alpha)\sin\alpha, \tag{2.27}$$

where $\lambda_{1(2)}$ is the photoelectron's mean free path in the sample (layer).

4. A semi-infinite sample partly covered by an overlayer of thickness t. For this

sample we have

$$N = D\lambda_1[1 - k + k\exp(-t/\lambda_2 \sin\alpha)]\sin\alpha, \quad (2.28)$$

where k is the part of the surface of the sample covered by the surface layer.
For the layer

$$N = D(1-k)[1 - \exp(-t/\lambda_2 \sin\alpha)]\sin\alpha. \quad (2.29)$$

5. The presence of a concentration gradient. So far in equation (2.23) it has been assumed that $n = \text{constant}$. As an example we shall consider the case $n = n_0 - bz$ with $n = 0$ at $z = a\lambda$, i.e. $n = n_0(1 - z/s\lambda)$. If λ does not depend on z, for the sample [integrating (2.24) from zero to $s\lambda$] we have

$$N = D/nn_0\lambda \sin\alpha\{1 + \sin\alpha/s[\exp(-s/\sin\alpha) - 1]\},$$
$$N \to O \quad \text{if} \quad s \to O, \quad N \to D/nn_0\lambda \quad \text{if} \quad s \to \infty. \quad (2.30)$$

In principle it is easy to estimate N for other types of $n = f(z)$ (see, for instance [34]). So,

$$C = (N/D)n. \quad (2.31)$$

Taking account of (2.1), (2.7), (2.20), and (2.31), it is easy to obtain equations for the intensity of X-ray photoelectron lines. For example, for a semi-infinite sample

$$I_1 = \mathscr{T}_0 A_0 \Omega_0 f(E_{\text{kin}}) F(E_{\text{kin}})$$
$$\times K \frac{\sigma_{nl}}{4\pi}[1 + \tfrac{1}{2}\beta(\tfrac{3}{2}\sin^2\theta - 1)]n\lambda(\alpha \geqslant \alpha_{\text{min}}),$$
$$I'_1 = I_1 \sin\alpha/\sin\alpha_{\text{min}} \qquad (\alpha \leqslant \alpha_{\text{min}}). \quad (2.32)$$

For a film of thickness t

$$I_2 = I_1[1 - \exp(-t/\lambda_2 \sin\alpha)] \qquad (\alpha \geqslant \alpha_{\text{min}}),$$
$$I'_2 = I_2 \sin\alpha/\sin\alpha_{\text{min}} \qquad (\alpha \leqslant \alpha_{\text{min}}). \quad (2.33)$$

For a semi-infinite sample covered with a continuous film of thickness t:

$$I_3 = I_1 \exp(-t/\lambda_2 \sin\alpha) \qquad (\alpha \geqslant \alpha_{\text{min}}),$$
$$I'_3 = I_3 \sin\alpha/\sin\alpha_{\text{min}} \qquad (\alpha \leqslant \alpha_{\text{min}}). \quad (2.34)$$

Dependence I on α as predicted in theory is presented in Fig. (2.1) [1]. In the case of a semi-infinite sample (Fig. 2.1a) the effective area is constant at $\alpha < \alpha_{\text{min}}$ (2.4), and in accordance with (2.25) and (2.32) I grows as $\sin\alpha$. At $\alpha > \alpha_{\text{min}}$ the value I does not depend on α (with the exception of the field of small angles Φ, see above). The surface layer of thickness t (Fig. 2.1b) and the semi-infinite sample under a layer of thickness t (Fig. 2.1b) at $\alpha > \alpha_{\text{min}}$ give intensity values for which the sum is roughly constant and depends little on α. Experimental data qualitatively accord with these results (Fig. 2.2 [1]) but a more detailed analysis shows that the theoretical equations are derived at on making a number of assumptions (for instance, an absolutely smooth surface, absence of elastic scattering, etc.) which it is difficult to achieve in practice. Besides, the comparatively small precision of

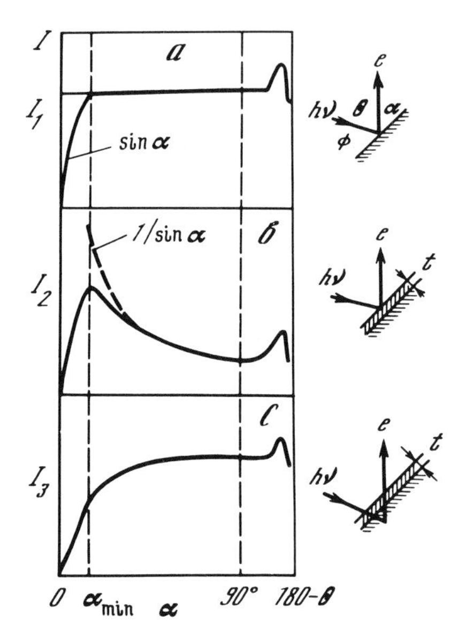

Figure 2.1. Theoretical dependence I_1 on angle α.

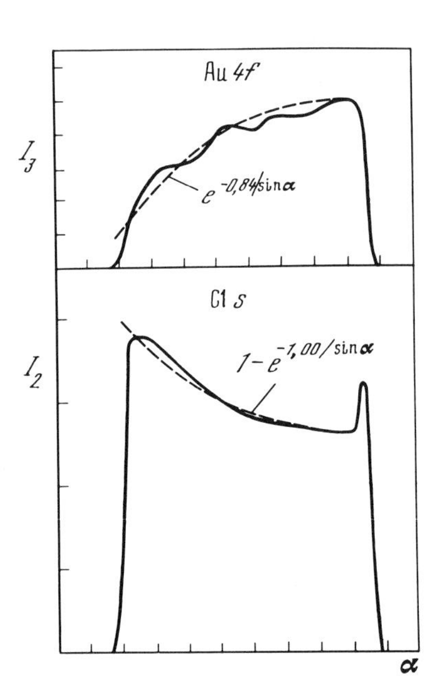

Figure 2.2. Dependence of experimental intensities of Au 4f and C 1s lines (continuous curve) on angle α for monocrystal Au covered by a carbon-containing film. The dashed curve presents the experimental data in form (2.34) for Au 4f and in form (2.33) for C 1s.

experimental data makes it possible to approximate experimental data by various sets of t/α depending on the equation used (2.33), (2.34) or their ratio [34] (see next subsection).

In conclusion let us dwell briefly on the influence of the sample's roughness on the intensity of the X-ray photoelectron lines. In the case of a semi-infinite sample without a contamination surface layer, the intensity does not depend on roughness, since the projection of any surface at the spectrometer's slot is $A_0/\sin\alpha$ at $\alpha > \alpha_{min}$. But if we study ratios of intensities from a substrate covered by a film of adsorbate to the signal intensity of the adsorbate, then along with dependence on α there must be a sensitivity to the profile of the surface [4]. Let us study the simplest example of a sample—metallic aluminum covered by a film of Al_2O_3. The sample was polished in such a way that it developed micro-grooves in a particular direction. The ratio of intensities Al 2p (oxide)/Al 2p (metal) was measured for two directions: parallel to the grooves (1) and perpendicular to them (2), depending on angle α between the surface and the photoelectron direction (Fig. 2.3) [4]. It turned out that in the first case the measured ratio of intensities was noticeably greater than in the second. The explanation is that in the second case the exit angle of photoelectrons α' on the sides of the grooves is noticeably bigger than the average angle α between the surface and the photoelectrons direction, this enhancing the strengthening of the signal Al 2p (metal) (see (2.34)). The theoretical results accord well with the experimental ones.

Let us consider yet another example (Fig. 2.4) [4]. Here the ratio of intensities Al 2p (oxide)/Al 2p (metal) for a sample with a profile as indicated in Fig. (2.4) was studied. Intensities of photoelectrons were measured in two directions relative to α: α^- and α^+ (see Fig. 2.4). When the electron exits in the direction α^- the measured ratio should behave in a qualitatively similar manner to the ideally smooth sample, i.e. the ratio should increase with decreasing α^-. Two processes should be observed during the decrease of α^+: on the one hand, the growth of the measured ratio according to (2.34), and, on the other hand, at angles close to 10° the emission should be only from the side walls of the lattice, this complying with angle $\alpha' \sim 90°$ and increasing the intensity of the metal line (see (2.34)).

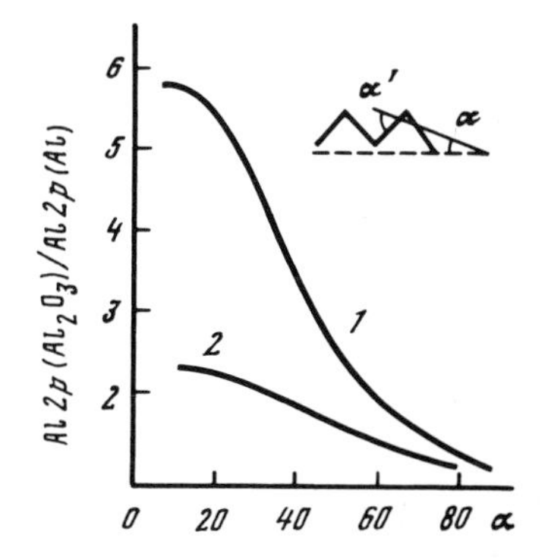

Figure 2.3. Theoretical dependence of intensities of Al 2p line on metal and oxide film on angle α when photoelectrons exit perpendicular (2) and parallel (1) to the direction of the grooves on the surface.

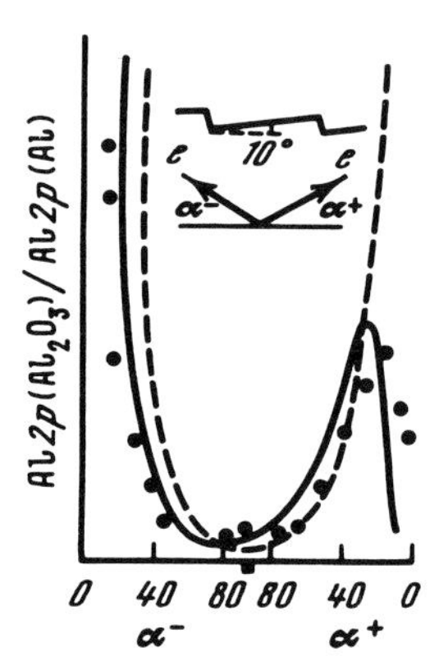

Figure 2.4. Dependence of intensities of Al 2p line on metal and oxide film on angles α^- and α^+. Continuous line—theory for the profile sample. Dashed line–theory for smooth sample. Dots show experimental data for profile sample.

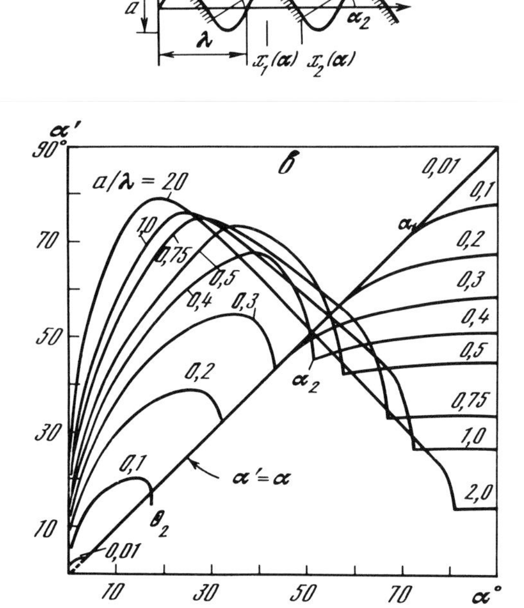

Figure 2.5. Dependence of (a) surface profile and (b) the interconnections α and α' on a/λ.

Consequently, the measure ratio has its maximum in the field α^+. Experimental data accord well with this computation [4].

It is also necessary to note that in the event of macroscopic surface roughnesses it is also necessary to take into account the screening of X-ray radiation. If the roughnesses are small, it is enough to take into account the changes in the mean exit angle as a result of the photoelectrons screening. As an example [1] let us consider a sinusoidal surface (Fig. 2.5a) with a period λ and amplitude α. Depending on the ratio a/λ and the angle α the mean exit angle of the photoelectron can be both greater or smaller than angle α. As a result of this, the effective thickness of the photoelectron's escape in the studied sample ($\lambda \sin \alpha'$) can be noticeably greater than in the case of the smooth sample ($\lambda \sin \alpha$) at small values of α (Fig. 2.5b). At usual exit angles of photoelectrons ($\alpha \sim 45°$) the ratio $\sin \alpha'/\sin \alpha$ is close to unity and, consequently, the usual evaluation of the escape depth $\lambda \sin \alpha$ is sufficiently precise.

At present the number of papers devoted to this extremely important problem—influence of roughness of intensity—is compartively small (see [35, 36]), although roughness, as it is seen from the previous examples, can substantially influence the results of surface composition analysis.

2.2. Mean free path of photoelectrons in a solid

As a photoelectron passes through a solid it can experience elastic and inelastic interactions connected with the ionization or excitation of valence and inner electrons, as well as the excitation of bulk and surface plasmons. If the cross-section of the electron's inelastic and elastic scattering by the atom is designated by σ_n and σ_e and the effect of the atomic interaction in a solid is ignored, then the mean free paths λ_n and λ_e are equal to

$$\lambda_n = M/(\textstyle\sum(i)\sigma_n^i)\rho \mathrm{N}, \quad \lambda_e = M/(\textstyle\sum(i)\sigma_e^i)\rho \mathrm{N}, \tag{2.35}$$

where M is the molecular or atomic mass, σ^i is the corresponding cross-section of atoms entering the molecule (solid), ρ is density and N is the Avogadro number.

The λ values can also be expressed in units of monolayers or in $\mathrm{mg\,m^{-3}}$ (λ_d) [37]:

$$\lambda_m = \lambda/\alpha, \quad \lambda_d = 10^{-3}\rho\lambda. \tag{2.36}$$

Here $a^3 = (M/\rho n\mathrm{N}) \times 10^{24}$, with λ expressed in 10^{-7} cm and ρ in $\mathrm{kg\,m^{-3}}$, while n is the number of atoms in the molecule.

The total length of the λ_t mean free path is

$$1/\lambda_t = 1/\lambda_n + 1/\lambda_e \tag{2.37}$$

Let us first study the values λ_n. Since the effects of elastic scattering have been ignored until recently [38–41] in XPS studies, the values λ appearing in equations in the previous section are identified with λ_n. The following three methods of determining λ_n are now mainly used. Some less common methods are described in [42]. See also [43, 44].

The method of changing the film thickness [45] is based on the deposition of the film on the substrate. The thickness t of the film is determined, for instance by measuring the increase in mass (by means of a quartz resonator). According to the Equation (2.34) we have

$$\lambda = t/\sin \alpha \ln(I_\infty/I_3), \tag{2.38}$$

where $I_\infty = I_1$ – intensity of substrate without the film [see (2.34)]. Determining the ratio I_∞/I_2 at various values of t also makes it possible to find λ; in particular, on the basis of (2.33) we have

$$\lambda = t/\sin\alpha[\ln I_\infty/(I_\infty - I_2)]. \tag{2.39}$$

The angular dependence method is also based on Equations (2.38) and (2.39), but in this case the film thickness t remains constant while α changes. Usually λ is found graphically or by the least squares method with the help of Equations (2.38) and (2.39) [34].

The third method [46, 47] is based on measuring the relative intensities of X-ray photoelectron lines from two samples. If some less substantial multipliers are ignored, on the basis of (2.32) for the ratio of intensities from two samples (1 and 2) we have

$$I(1)/I(2) = \lambda_1\sigma_1 f_1(E_{\text{kin}})/\lambda_2\sigma_2 f_2(E_{\text{kin}}) \tag{2.40}$$

λ_1 can be found from this expression if the σ_1/σ_2 theoretical or empirical evaluations are used and if λ_2 is known.

Let us dwell briefly on the theoretical calculations λ_n before going on to an analysis of the experimental values. The simplest variant of estimating σ_n is to ignore the atomic interaction in the solid and use σ_i values for free atoms [see (2.35)]. Extensive calculations of σ_n values in the Born approximation for virtually all free atoms are presented in [48–50]. Also studied in these references is the interconnection of photoionization cross-sections of A *nl* shells with the inelastic electron scattering cross-section. The estimated σ_n values usually agree comparatively well with experimental data for free atoms. It should also be noted that σ_n calculations can be done by the semi-empirical equation proposed in [51].

λ_n values that take account of the specificity of the electron structure in the compound under consideration were calculated in references [42, 52–56].

Table 2.3
Values a and b in Equation (2.41)

Z	a	$-b$	Z	a	$-b$	Z	a	$-b$	Z	a	$-b$
3	4.97	1.23	25	19.7	2.87	41	14.2	2.37	57	9.47	1.60
4	11.7	2.32	26	21.1	3.00	42	16.9	2.56	72	11.4	2.14
5	10.8	2.70	27	22.8	3.14	43	18.5	2.72	73	13.8	2.40
6	18.3	2.95	28	23.7	3.21	44	19.9	2.85	74	16.2	2.58
7	4.66	1.05	29	23.6	3.21	45	20.6	2.94	75	18.3	2.71
11	4.84	1.25	30	21.1	3.10	46	20.6	2.98	76	19.1	2.86
12	7.74	1.77	31	17.3	2.00	47	19.5	2.95	77	19.4	2.95
13	10.2	2.16	32	14.0	2.22	48	17.4	2.81	78	19.4	2.96
14	10.7	2.19	33	13.7	1.69	69	14.0	1.80	79	19.6	2.95
19	3.55	0.820	34	11.9	2.25	50	12.9	2.04	80	16.6	2.71
20	6.67	1.48	35	9.58	2.46	51	12.4	2.18	81	13.4	1.62
21	9.92	1.94	37	3.51	0.71	52	12.0	2.22	82	13.2	1.97
22	13.2	2.28	38	5.55	1.24	53	10.0	2.08	83	12.0	2.02
23	16.5	2.54	39	8.52	1.71	55	3.20	0.57			
24	19.0	2.74	40	11.5	2.16	56	5.59	1.15			

Note: Z = order number of element.

Admittedly, as a rule valence electrons were considered in the approximation of free electron gas. The most complete tables of values for λ are contained in [53], in which the following equation is used:

$$\lambda = E_{kin}/a(\ln E_{kin} + b), \tag{2.41}$$

where E_{kin} is the kinetic energy of the photoelectron, a and b are constants given in Table 2.3, and λ is in Å. Equation (2.41) holds when $E_{kin} > 200$ eV. According to estimates [53, 54] the main contribution to the value σ_n is made by valence electron scattering. For instance, the contribution of inner electrons amounts to 14, 12, or 9 per cent to the total value σ_n for metallic aluminum at $E_{kin} = 10000$, 1000, and 500 eV, respectively [52]. According to estimates [53], λ usually decreases for elements along the Periodic Table and increases in groups from the top downwards. For example, at $E_{kin} = 1000$ eV, $\lambda = 3.65$, 2.51, and 1.96 nm for Na, Mg, Al, and Si; and $\lambda = 2.51$, 2.76, 3.17, and 3.10 nm for Mg, Ca, Sr, Ba. The value λ [53] increases with E_{kin} roughly according to the law $\lambda \sim E_{kin}^{0.77}$ for E_{kin} values around 700 eV [37]. The expression $\lambda \sim E_{kin}^{0.65 \div 0.80}$ [57] has been obtained for a number of compounds and metals at E energies of 200 eV $< E_{kin} <$ 10 keV and $\lambda \sim E_{kin}^{0.53 \div 0.80}$ at 150 eV $< E_{kin} <$ 4 keV [58]. According to the theoretical

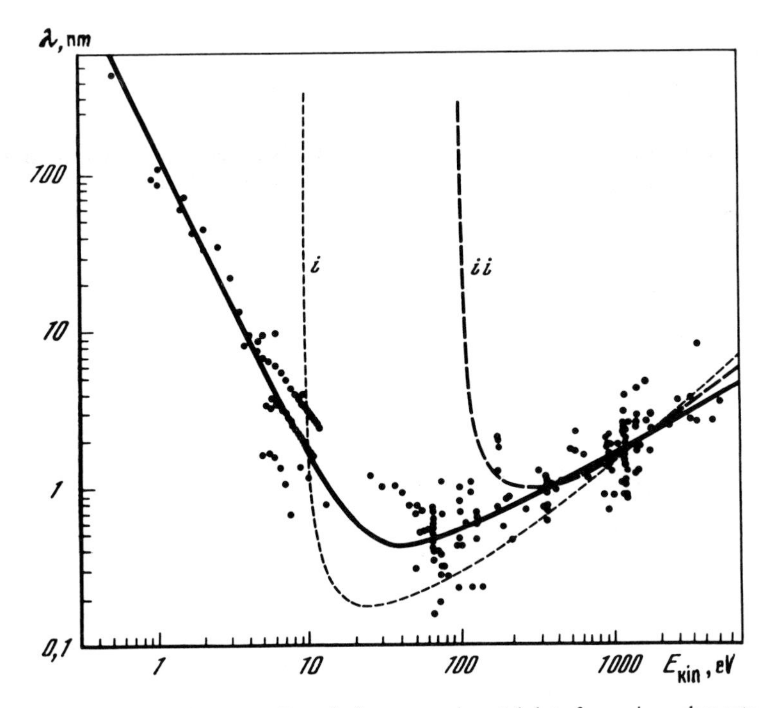

Figure 2.6. Dependence of λ on E_{kin}. Dots indicate experimental data for various elements. Continuous line—average statistical 'universal' curve. Line i corresponds to the mean values of a and b in (2.41), while line ii corresponds to the values a and b in (2.41); these values agree best with the experiment at $E_{kin} > 150$ eV.

paper [59], at small values of E_{kin} the following ratio is true:

$$\lambda \sim E_{kin}^{-2}, \tag{2.42}$$

i.e. λ decreases with the growth of E_{kin} at small values of E_{kin}. This means that according to this theory a minimum must be observed at some values of E_{kin}. This minimum really manifests itself both for experimental values of λ [37] and for λ_n for free atoms (with $E_{kin} \sim 50$ eV).

Several hundred values of λ at different E_{kin} of various elements have now been defined (see review [37] and some recent references [43, 44, 57, 60–67]). Experimental data confirm the existence of a minimum λ at $E_{kin} = 30 \div 100$ eV. Figure (2.6) [37] contains data for λ_n of various elements and inorganic compounds. The results show that there roughly exists a 'universal curve' [42] of λ_n versus E_{kin} (although the differences in the curves for elements and for inorganic compounds can be seen even on a logarithmic scale). Considering the large variability of experimental values of λ as determined by various authors and also the vagueness of theoretical assumptions used in the measurements (see below), the statistical approach [37] is quite appropriate at present when defining λ and its dependence on E_{kin}. The following equations to determine λ_i $(i = n, m, d)$ are suggested in the review [37] on the basis of a statistical processing of experimental material $\lambda_i (i = n, m, d)$:

$$\lambda_i = A_i/E_{kin}^2 + B_i E_{kin}^{1/2} \quad \text{(for all } E_{kin}), \tag{2.43}$$

$$\lambda_i = B_i E_{kin}^{1/2} \quad (E > 150\,\text{eV}), \tag{2.44}$$

where A and B are given in Table (2.4) [parameter B is the same for Equations (2.43) and (2.44)].

Table (2.4) contains the number N of λ values used in the statistical processing

Table 2.4
Parameters A_i and B_i in Equations (2.43), (2.44)

		λ_n, Å		λ_m		λ_d, mg m^{-2}	
Material	N	B_n	δ	B_m	δ	B_d	δ
Au	22	0.54	1.30	0.209	1.30	1.036	1.30
Elements	102	0.54	1.47	0.210	1.38	0.336	2.23
Inorganic compounds	52	0.96	1.70	0.365	1.47	0.298	1.56
Organic compounds	57	0.87	2.12			0.110	2.10
Adsorbed gases	3	0.64	1.98				
		$A_n \times 10^{-3}$	δ	A_m	δ	A_d	δ
Au	82	1.77	1.48	688	1.48	3400	1.48
Elements	215	1.43	1.57	538	1.59	2220	2.44
Inorganic compounds	59	6.41	1.70	2170	1.48	3630	1.53
Organic compounds	61	0.031	2.07			49	2.05

Table 2.5
Dependence of t/λ values on determination method

Method	[68]	[69]	[70]	[34]
I	0.67	0.43	0.41	1.16
II	0.28	0.33	0.02	0.37
III	−0.02	0.09	−0.36	−0.11

and the mean standard deviation δ for the value $\ln\lambda$. Parameters A and B are different for elements, inorganic compounds, and organic compounds. The values λ estimated according to (2.43) and (2.44) agree better with existing experimental data than those estimated according to (2.41) [37]. It should be noted that reference [66] proposes a linear dependence of λ on E_{kin} for a large number of halogenides.

It must be borne in mind that because of the imperfection of the methods used, the experimental values λ are determined with a comparatively low degree of precision. For instance, when using the method of changes in film thickness, the thicknesses of the film being small, it is difficult to control the uniformity of the overlayer and determine thickness with sufficient precision. The method based on measuring the angular dependence of intensity [34] is especially sensitive to all sorts of deviations. Let us dwell on the result of an analysis of possible errors. Table 2.5 contains values of t/λ determined on the basis of experimental data in references [34, 68–70]. The values t/λ were determined on the basis of angular dependence by three methods corresponding to Equations (2.39) (I), (2.38) (II), and the equation that can be obtained by dividing (2.33) by the expression (2.34) (III). The negative values t/λ are an artefact of the mathematical processing used [34] and attention should be paid only to the course of changes in λ values depending on the method. The results obtained point to the decrease of the values of t/λ found during the transition I→II→III, despite the use of the same experimental data. Moreover, the values themselves turned out to be dependent on the range of angles α used for determining t/λ. The values of t/λ obtained may vary in individual studies. Attention should be given only to the dependence on method within the framework of the same study.

These facts show that the theoretical equations used to analyse experimental data on the basis of the angular dependence of intensities are not adequate. Reference [34] studied the influence of the pores in the overlayer on the substrate, the existence of a concentration gradient, and surface roughness on the value of t/λ, determined by simplified equations similar to (2.38) and (2.39), which ignore these factors. The analysis showed that in this case t/λ must indeed have various values for different methods and must depend on the range of α values used. According to this analysis, the observed dependence of t/λ values on the method and the mean value of α can be explained by the existence of roughnesses on the surface.

Reference [60] defines λ for cadmium arachidate (cadmium salt of the organic acid: CdAA) by the three methods studied above. By the method of measuring film thickness with a constant angle α a value of $\lambda = 5.1$ nm was obtained, with

$E_{kin} = 1402$ eV (Au 4f line). Using the same line, λ values from 12.0 to 4.2 nm with angle α changing from 10 to 55° were found according to Equation (2.34) by the method of changing angle α with constant film thickness t. This strong dependence on the angle is explained by the fact that the substrate was not fully covered by CdAA. If the part of the free substrate surface is designated K, then intensity from the substrate is presented by the equation

$$I = KI_{\infty} + (I - K)I_3, \tag{2.45}$$

where I_3 is given in Equation (2.34) and I_{∞} is the substrate's intensity without the overlayer.

Obviously, when α is small the contribution of I_{∞} to I is great, and according to (2.45) and (2.34) this must lead to a growth of λ. To lessen the dependence of λ on α, the authors [60] assumed that K amounts to about 1% for one sample and 4% for the other. In this case, with α changing from 20 to 55°, the range of λ amounts to 4.2–3.4 nm. When using the first method and taking into account K, the value of λ decreases from 5.1 to 4.1 nm. Besides the coefficient K, surface roughness (see previous section) also plays a substantial role when the seeming value of λ increases at small values of α.

When measuring relative intensities of CdAA, Au, and Ag samples, values from 4.0 to 5.7 nm were obtained for λ in CdAA, depending on the choice of parameters in Equation (2.40). So the results of the study [60] rather patently show that at present the value of λ cannot be determined experimentally with sufficient precision. More than that, the choice of the value K in (2.45) is quite subjective. Along with purely experimental difficulties, note should be made of two fundamental moments connected with λ values in equations for the intensity of X-ray photoelectron spectra.

Firstly, in connection with the dependence of the excitation of bulk and surface plasmons on the distance of the photoelectron from the surface, the value λ can also depend on this distance. Evaluations [71] show that this dependence is substantial only for distances of several tenths of nm from the surface: on the whole, changes in the probability of the excitation of bulk and surface plasmons compensate each other. The value λ begins to decrease at a distance of about 0.2–0.4 nm from the surface but does not drop to zero on the surface. The value λ is not equal to zero at a distance of several more tenths of nm, this distance being determined by the speed with which electron density decreases.

Secondly, the dependences cited above connecting λ and I were obtained without taking account of elastic scattering. But the cross-section of elastic scattering is approximately equal to the cross-section of inelastic scattering and, therefore, should noticeably influence the photoelectron's path in the solid. The intensity I, taking due account of elastic scattering, is calculated in [38–40, 72–76] using the Monte Carlo method. We shall dwell on the main results.

2.2.1. *Flat semi-infinite sample*

Estimates show that accounting for elastic scattering leads to an isotropization of the angular distribution of the photoelectrons, this formally being equivalent to a decrease in β (see 2.19). If the angle θ between the X-ray radiation and the exit of the photoelectrons is small, the experimental intensity is higher than that given by (2.19) which does not account for the elastic scattering of electrons. If θ is large, the intensity is smaller. This can lead to a increase or decrease in λ, if the expression (2.40) is used to determine it. At $\theta = 90°$ this effect is admittedly small.

2.2.2. *Sample covered with a film of* t *thickness*

As estimated, intensity I_3 of the substrate's line usually decreases as a result of elastic scattering, the relative effect increasing with t. But the opposite picture is observed at small values of the photoelectron exit angle α. When employing Equation (2.38) for various values of t and a fixed value of angles $\alpha > 60°$, the formal value λ is smaller, and for $\alpha < 60°$ is bigger than the true one. The effect amounts to about 20%. It shall be stressed that the estimated intensities of photoelectrons I with due account for elastic scattering and, consequently the experimental data can be presented with sufficient precision in the form of (2.34), but the value λ in the formal expression (2.34) (we will designated it λ_f) does not coincide with the value λ_n for the mean free path of electrons without inelastic collisions. Since

$$\lambda_f^{-1} = \lambda_n^{-1} + a\lambda_e^{-1},$$

where λ_e is the mean free path without elastic collisions and coefficient a takes into account the contribution of elastic collisions, $\lambda_f < \lambda_n$ is usually true. If (2.38) is used to determine λ at a fixed value of t and for various values of α, estimates show that the value λ will change with α. In particular the above studied results (from reference [60]) concerning the dependence of λ on α are explained, at least in part, by the effects of elastic scattering.

The value λ also is often estimated on the basis of the ratio of the intensity of the substrate I_3 (2.34) to the film intensity I_2 (2.33):

$$I_2/I_3 = R[1 - \exp(-t/\lambda \sin\alpha)] \exp(t/\lambda \sin\alpha), \tag{2.46}$$

where R is the ratio of intensity from the infinitely thick film to the intensity of the pure substrate.

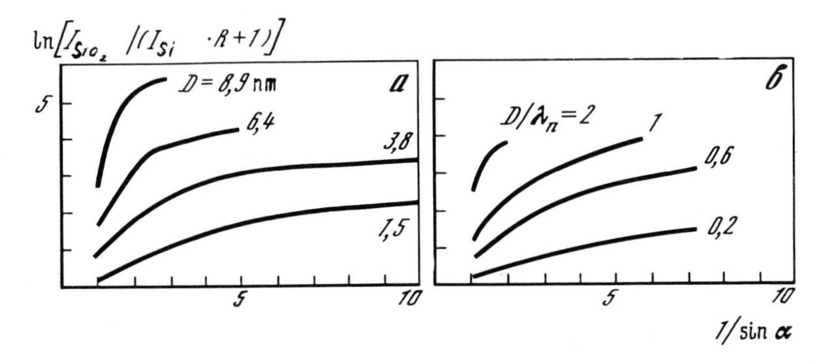

Figure 2.7. Dependence of $\ln[I_{SiO_2}/(I_{Si \times 0.672} + 1)]$ on exit angle θ for various thicknesses D of the SiO_2 film on Si. (a) Experimental results. (b) Calculation by the Monte-Carlo method; λ_n = mean free path without inelastic scattering.

The angular distribution of photoelectrons for a sample of silicon covered with oxide layers of various thickness t was obtained experimentally in [41]. Figure (2.7a) presents the dependence $\ln(I_2/I_3R+1)$ on $1/\sin\alpha$, taken from [34, 41]. The difference from the linear dependence that follows from the Equation (2.46) and also the dependence of λ on α were explained by the imperfection of the sample's surface. Figure (2.7b) presents the dependence $\ln(I_2/I_3R+1)$, I_2 and I_3 being estimated by the Monte Carlo method [39]. The qualitative agreement between Fig. (2.7a) and (2.7b) is quite obvious so the dependence observed in Fig. (2.7a) can also be explained by effects of elastic scattering.

The validity of using Equation (2.46) for determining t/λ, taking account of the limited size of the angular aperture $\Delta\alpha$ (see Fig. 1.4) in the spectrometer's energy analyser, was studied in reference [75]. Indeed, in this case the signal intensity from the film I_2 is expressed by the following equation rather than by expression (2.33):

$$I_2 = \int_{\alpha-\Delta\alpha/2}^{\alpha+\Delta\alpha/2} [1-\exp(-t/\lambda\sin\alpha)]\,\mathrm{d}\alpha$$

while the intensities ratio of the substrate and the film:

$$\frac{I_2}{I_3} = R\frac{\int_{\alpha-\Delta\alpha/2}^{\alpha+\Delta\alpha/2} [1-\exp(-t/\lambda\sin\alpha)]\,\mathrm{d}\alpha}{\int_{\alpha-\Delta\alpha/2}^{\alpha+\Delta\alpha/2} \exp(-t/\lambda\sin\alpha)\,\mathrm{d}\alpha}.$$

It was established that even at small values of $\Delta\alpha(=5^\circ-10^\circ)$ the above-cited values of I_2/I_3 at $\alpha<30^\circ$ substantially differ from the ratio (2.46). Here the deviations turn out to be in the same direction as those found when analysing the influence of roughnesses [34] and elastic scattering [38–40] on the angular distributions of photoelectrons (it should also be noted that references [38–40] took into account the combined influence both of the elastic scattering of electrons and the limited size of the analyser's angular aperture on the intensity of X-ray photoelectron spectra).

The distortions of angular distributions, caused by the imperfection of the sample surface, the limited size of the analyser's angular aperture or the elastic scattering of electrons, are qualitatively analogous: in all instances effective values of λ increase with the decrease of the angle α. In [39, 75] it is noted that all these deviations from the ideal situation, within the framework of which Equations (2.32)–(2.34) were obtained, are ultimately connected with the specificities of the trajectories of photoelectrons in the sample. In the ideal case the photoelectron's path in the sample depends only on the exit angle α and the depth at which it originated. In all other instances the paths of photoelectrons reaching the analyser are subordinated to a certain distribution depending either on characteristics of the sample surface and material, or on the experimental geometry.

So for all practical purposes, at present there is no reliable precise experimental data on λ values. Nevertheless, the order of λ values, and the qualitative dependence of λ on E_{kin} and on the element's place in the periodic system are known sufficiently well. Considering the vagueness and comparatively low accuracy of experimental values, we recommend at present use of the results of statistical processing, cited in references [37] (see above) to assess values. In those cases when dependence of λ on the element's place in the periodic system plays a substantial role, use can be made of estimates [48, 50, 53].

2.3. Quantitative surface analysis of solids

It follows from the previous section that the length of the photoelectron mean free path in a solid at $E_{kin} \sim 1000\,\text{eV}$ amounts to about 1.5–4.0 nm. Since the intensity of the X-ray photoelectron line falls according to the exponential law $I \sim \exp(-z/\lambda)$ depending on the studied volume's distance z to the surface, then according to (2.33) the contribution of a surface layer of λ, 2λ, and 3λ thickness amounts to 63, 87, and 95%, respectively, of the X-ray photoelectron line's total intensity. Hence, X-ray photoelectron spectra carry information on the sample surface layer of the order of 1.5–4.0 nm. The method's surface sensitivity can be increased by reducing the angle* α or by reducing E_{kin} by selecting a level with a greater binding energy and an X-ray quantum with a lesser energy. These questions are considered below. Since the X-ray photoelectron line's intensity is in direct proportion to concentration, X-ray photoelectron spectra make it possible to judge the quantitative composition of superthin surface films.

According to (2.32), with $\alpha > \alpha_{min}$ and $f(E_{kin}) \sim E_{kin}^{-1} = E^{-1}$ for the intensity ratio I_1/I_2, we have†

$$I_1/I_2 = \frac{n_1\sigma_1[1 + \frac{1}{2}\beta_1(\frac{3}{2}\sin^2\theta - 1)]\lambda_1 K_1 E_2}{n_2\sigma_2[1 + \frac{1}{2}\beta_2(\frac{3}{2}\sin^2\theta - 1)]\lambda_2 K_2 E_1} \tag{2.47}$$

Usually $\theta = 90°$ and β varies between 1 and 2 (see Table 2.2) and for this reason the maximum error when ignoring the multiplier $[1 + \frac{1}{2}\beta(\frac{3}{2}\sin^2\theta - 1)]$ does not exceed 20%. In the vast majority of instances the value K amounts to 0.9–0.8, and for this reason the difference between K_1 and K_2 can also be ignored in the first approximation. For λ values one can approximately accept $\lambda \sim \sqrt{E_{kin}}$ [see (2.44)]. With due account for these simplified assumptions we have

$$I_1/I_2 = n_1\sigma_1\sqrt{E_2}/n_2\sigma_2\sqrt{E_1}. \tag{2.48}$$

Since in real samples there is always a contamination layer on the surface with a thickness of the order of λ, the more precise expression for I_1/I_2 taking into account Equation (2.34) appears as

$$I_1/I_2 = \exp(-t/\lambda_1 + t/\lambda_2) n_1\sigma_1\sqrt{E_2}/n_2\sigma_2\sqrt{E_1} \tag{2.49}$$

---

*Depth of studied layer is proportional to $\sin\alpha$ (2.33).

†$F(E_{kin})$ should be constant since it is assumed that spectrum scanning is by retarding field and electrons with a constant E_{kin} enter the detector. Indices 1 and 2 are related to samples 1 and 2.

where t is the thickness of the contamination layer, and λ_1 and λ_2 are the lengths of the mean free path in this layer at energies E_1 and E_2. The value of t can vary within a broad range but usually it is of the order of about 1 nm, i.e. approximately $\frac{1}{2}\lambda \div \lambda$ at $E_{kin} = 500 \div 1000$ eV.

Let us assume that $t \sim \lambda_1\lambda_2/(\lambda_1 + \lambda_2)$. Then on the basis of (2.49) we get $I_1/I_2 = Cn_1\sigma_1/n_2\sigma_2$,

$$C = \sqrt{E_2/E_1}\exp[(\sqrt{E_1} - \sqrt{E_2})/(\sqrt{E_1} + \sqrt{E_2})] \quad (2.50)$$

Analysis of the composition is usually made at $E_{kin} = 500 \div 1400$ eV. Assuming that $E_2/E_1 \sim 1.5$ we have $C \sim 1.1$. Even if we assume $E_2/E_1 \sim 2.5$ we have $C \sim 1.25$. The relative values σ_1/σ_2 change within a rather broad range (see Table 2.1) and for this reason it should not be assumed that $\sigma_1/\sigma_2 \sim 1$ should not be assumed. Hence, the simplest variant of quantitative analysis on the basis of X-ray photoelectron spectra is founded on the equation

$$n_1/n_2 = (I_1/\sigma_1)/(I_2/\sigma_2). \quad (2.51)$$

The ratio (2.51) is really performed with an even greater precision than could have been supposed considering the nature of the assumptions made when this ratio was derived. Figure (2.8) [13, 14] shows quite a high degree of agreement between the ratios I_1/I_2 and σ_1/σ_2 ($n_1/n_2 = 1$). Equation (2.51) can be used with the greater confidence the closer the ratio E_2/E_1 is to unity. It is desirable for the

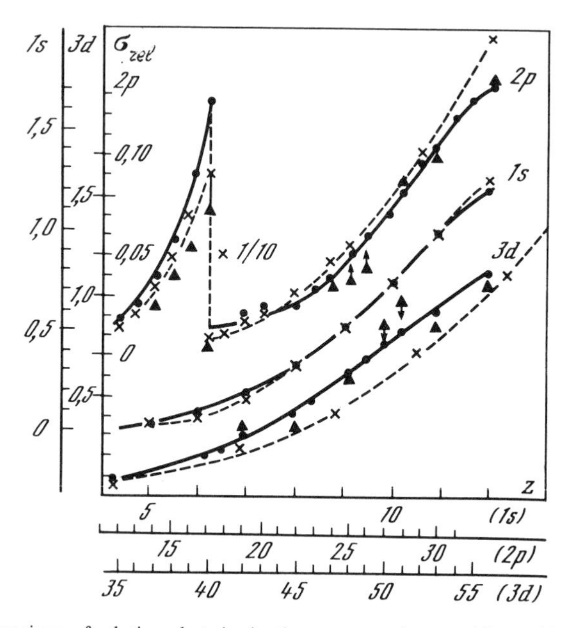

Figure 2.8. Comparison of relative photoionization cross-sections and intensities of X-ray photoelectron lines (referred per one electron). Data for Na ls taken as unity. Dots and triangles are experimental data on relative intensities, crosses are theoretical values for relative photoionization cross-sections.

ratio E_2/E_1 to be not greater than 2, in which case the error for n_1/n_2 usually does not exceed 20–30% (assuming that σ_1/σ_2 is known).

The high degree of agreement between I_1/I_2 and σ_1/σ_2 is not achieved at $E_2/E_1 \sim 4$ because in this case the value of the exponential member in (2.49) substantially depends on t. As an example, Fig. (2.9) presents the dependence of the ratio $I(\text{Zn 3p})/I(\text{Zn 2p})$, where $E_1/E_2 \sim 4$, on the intensity $I(\text{Zn 2p})$ of the sample of metallic zinc. $I(\text{Zn 2p})$ increases with the decrease of the contamination layer thickness t that is achieved by ion sputtering. Although the absolute values $I(\text{Zn 3p})$ also increase, the ratio $I(\text{Zn 3p})/I(\text{Zn 2p})$ decreases because in (2.49) $\lambda_1 \gg \lambda_2 (\lambda_1 \sim 2\lambda_2)$ and the value of the exponential member noticeably decreases with the decrease of t. The exceptionally great range of measured [26] values of $I(\text{Na 2s})/I(\text{Na 1s})$ (from 0.036 to 0.081 in the case of the excitation of spectra by the Mg K_α line, where $E_1/E_2 = 6.5$, and from 0.044 to 0.069 in the case of the excitation of spectra by the Al K_α line, where $E_1/E_2 \sim 3.4$) in various compounds is connected both with the influence of the exponential member in expression (2.49) and with the change in the value p in the relationship $\lambda \sim E^p$ for various compounds.

So the value of the photoionization cross-section is a value that, in the main, determines the dependence of relative intensities of X-ray photoelectron lines (given equal concentrations). The theoretical values of σ can be used for the practical application of ratio (2.51). The comparatively high degree of precision of the calculated σ follows both from a comparison of the estimated total photoionization cross-sections (kbarn) of atoms for the Al K_α line with experimental ones [16] and from the rather good agreement of the theoretical and 'experimental' values of σ_1/σ_2 determined in [66, 26, 77, 78] on the basis of ratio (2.47), in which I_1/I_2 is taken from the experiment while the other parameters are evaluated.

For example, theoretical values were taken for σ, the dependence $\sim E^p$ was used for λ, the value K was ignored, etc. While relative 'experimental' values of σ

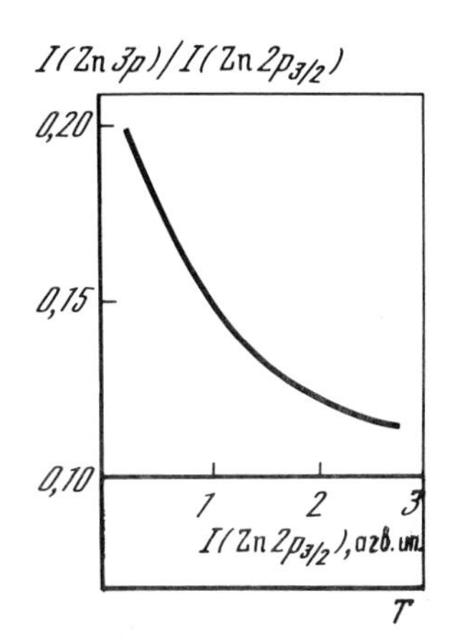

Figure 2.9. Dependence of $I(\text{Zn 3p})/I(\text{Zn 2p}_{3/2})$ on absolute intensity $I(\text{Zn 2p}_{3/2})$ in the process of removing contamination surface layer (T = time of ion sputtering).

are not necessarily more accurate than the theoretical ones, they are useful for identifying some theoretical ratios of σ_1/σ_2 which do not give a correct description of the observed I_1/I_2. This is true of, for example, $\sigma 2s/\sigma 2p$ elements K and Ca [17, 26].

A comparison of calculated and experimental total photionization cross-sections (in kbarn) for the line Al K_α is cited below, as an example [16].

Atom	Be	C	N	O	F	Ne	Al	S	Cl	Ar
Calculation	2.76	14.4	25.7	42.2	63.9	91.3	18.3	45.3	58.7	74.8
Experiment	2.67	14.3	25.8	42.2	64.3	91.2	18.1	46.1	60.1	77.0

When determining the relative 'experimental' values σ_{nl}, an important part is played by the method of determining I_1/I_2 [78]. By its physical meaning the calculation of photionization cross-sections presupposes the determination of the integral intensity I_1 (area between the line profile and background level). The line profile is set by the equation

$$I(E) = \{\cos[\tfrac{1}{2}\pi\alpha + (1-\alpha)\arc\tan E/\gamma]\}/[E^2 + \gamma^2]^{(1-\alpha)/2} \tag{2.52}$$

where 2γ is the line width at $I = \frac{1}{2}I_{max}$; and α is the asymmetry parameter, which equals 0 for insulators and diminishes within the limits from 0 to 0.3 for metals (α grows with the increase of electron density near the Fermi level [79]). Two methods of determining I are studied in [78]: measuring an area with a constant ΔE, for instance 5 eV, for I lines (method A); and measuring line area with a constant range γ, for instance $\pm 4\gamma$ (method B). Table (2.6) [78] shows how the line's area changes in so doing. When using the theoretical values σ, β and the dependence of λ on E [see (2.41)], the calculated values of I_1/I_2 agree better with the experimental values using method B. This result is obvious because method B ensures that due account is taken of the same part of the line's total intensity.

The method of graduation curves [16, 17, 24] is used to obtain the most precise values of n_1/n_2 (with a precision of 5–10%). The dependence of the ratio I_1/I_2 on n_1/n_2 for a number of compounds is determined within the framework of this method (Fig. 2.10 [16]). This makes it possible to take into account with precision ratios connected with σ and β, almost precisely ratios connected with the value K, most ratios of λ in the correlation (2.47), and the value of the expontial member in (2.49). The method makes it possible to rule out chance deviations and to average matrix effects connected with values of λ and the

Table 2.6
Relative intensities of In[a] lines

Level	γ, eV	Method A	Method B	Calculation
4d	0.7	0.34	0.35	0.36
4s	2.4	0.05	0.08	0.10
$3d_{5/2}$	0.6	1	1	1
$3p_{3/2}$	1.8	0.26	0.34	0.33
$3p_{1/2}$	1.8	0.11	0.14	0.14

[a]Spectra were excited by the Mg K_α line.

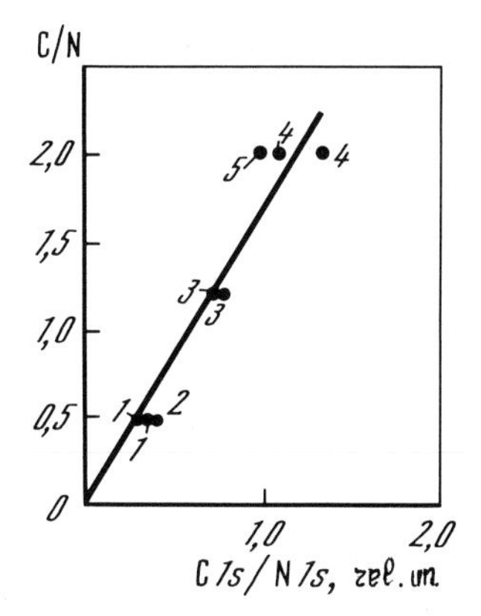

Figure 2.10. Graduation curve for determining atomic ratios C/N. Figures designate various compounds (see [16]).

exponential member. With atomic ratios A/B changing within limits to 8 it has not been possible to date to discover substantial deviations from linearity in the relationship between the ratios I_A/I_B and A/B. Generally speaking, matrix effects should be expected in the event of A and B being small, if the values λ_1 and λ_2 for the matrixes 1 and 2 differ substantially, and if E_{kin} for A and B also differ. But such effects have not been discovered to date, and for this reason relative line intensities for any A/B ratio and any matrix are used for analysis. Experimental values of relative intensities are given in Table (2.1). When the graduation curve method is used it does not matter how integral intensity is measured. It is important to use the same measurement method when plotting the graduation and when determining n_1/n_2 on the basis of I_1/I_2. The area within an experimental line profile and a background line was measured in references [16, 17, 24]. This yields a value of I_1 proportional to σ_1 [16, 17, 77, 24].

It is necessary to note that when plotting graduation curves one must take substances whose surface composition is not distorted by hydrolysis, oxidation, etc. The X-ray photoelectron lines used for plotting a graduation curve should not be distorted by the imposition of other lines. For example, when determining the ratio of line C 1s/N 1s (see Fig. 2.10), use was made of substances in which the C 1s line does not coincide in energy with the C 1s line from the hydrocarbon contamination layer.

Mixtures of substances should not be used for graduation because the method is sensitive only to the surface layer: as a result line intensity changes when one of the components is more finely ground or when the surface is made up of particles of one component of the mixture. Reference [80] shows, in particular, that the adding of graphite to a mixture of lead salts and MoQ_3 results in a substantial change in the slope of I(Pb 4f)/I(Mo 3d) graduation curve depending on the Pb/Mo ratio for various lead salts.

A certain caution is necessary also when using values of relative intensities obtained using another spectrometer (even of the same type as the one used for analysis). Although the data of various authors on relative intensities given in

Table 2.1 agree comparatively well with one another,* the result of a Round Robin [81] of relative intensities obtained by various authors showed the possibility of an exceptionally great scattering of experimental values. This scattering is explained in part by malfunctions in instruments but mainly, it appears, is due to the non-identical experimental conditions used by the various researchers, even if these conditions were set in advance.

Thus, the result of the latest international Round Robin [82] produced a comparatively good coincidence of results concerning relative line intensities: in most instances relative intensities coincided within 10% [with due account for differences in the $f(E_{\text{kin}})$ values for various spectrometers]. The results of this check [82] show that in principle the graduation curves obtained on one spectrometer can be recalculated for use in another spectrometer.

It was assumed above that all the investigated elements are evenly distributed in the surface layer of the studied sample. In this case $n = \text{const}$ and $I \sim n\lambda$. Let us study instances [83–85] when there is a gradient of concentrations, that is $n = f(z, s_i)$ and $I \sim \lambda f(s_i, n_i)$, where the parameters s_i and n_i characterize the concentration gradient. Let us study [83] the three most important cases (Fig. 2.11):

(I)
$$n = n_0 - (n_0 - n_1)z/s\lambda, \quad 0 \leqslant z \leqslant s\lambda$$
$$n = n_1 < n_0, \quad z \geqslant s\lambda; \tag{2.53}$$

(II)
$$n = n_1 + z(n_0 - n_1)s\lambda, \quad 0 \leqslant z \leqslant s\lambda,$$
$$n = n_0 > n_1, \quad z \geqslant s\lambda; \tag{2.54}$$

(III)
$$n = n_0 \exp(-z/s\lambda), \qquad 0 \leqslant z < \infty. \tag{2.55}$$

On integrating the expression $1/\lambda \sin\alpha \int_0^\infty n \exp(-z/\lambda \sin\alpha) \mathrm{d}z$ we will get the

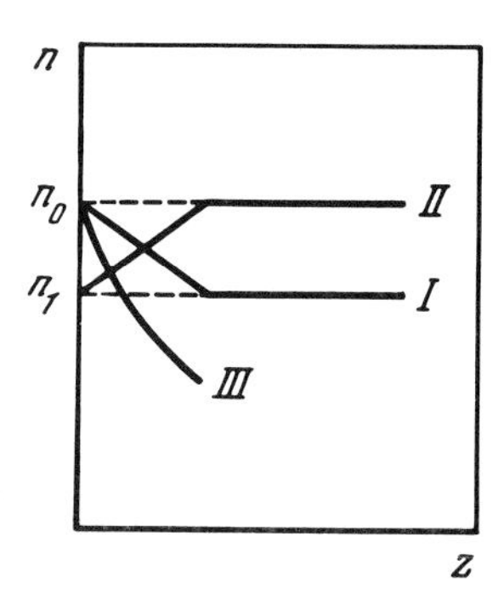

Figure 2.11. The dependence of concentration n on distances z from the surface of the sample.

*The discrepancy is partly caused by the circumstance that in [25] $I \sim f(E_{\text{kin}}) \sim E_{\text{kin}}$, while in [24, 26] $I \sim f(E_{\text{kin}}) \sim E_{\text{kin}}^{-1}$. It should also be remembered that the thickness of the contamination layer depends on the vacuum in the spectrometer and as a result it varies, even for spectrometers of the same type.

following values of $f(s, n_i)$:

(I) $f(s, n_i) = (n_0 - n_1)\{1 + (\sin\alpha/s)[\exp(-s/\sin\alpha) - 1]\} + n_1$ (2.56)

(II) $f(s, n_i) = (n_0 - n_1)(\sin\alpha/s)[1 - \exp(-s/\sin\alpha)] + n_1,$

(III) $f(s, n_0) = n_0 s/(s + \sin\alpha).$

It can be shown that the concentrations n_{av}, obtained on the basis of experimental intensity values without due account of the variety $f(s, n_i)$, that is

$$I \sim f(s, n_i) = n_{av} \tag{2.57}$$

in a sufficiently wide range of s are extremely close to the values n_{av} characterizing the average concentration in a surface layer x thick:

$$\bar{n}_{av} = (1/x\lambda)\int_0^{x\lambda} n(z)\mathrm{d}z. \tag{2.58}$$

It is shown in [83] that the approximate equality

$$f(s, n_i) = n_{av} \approx \bar{n}_{av} \tag{2.59}$$

occurs at $x \sim 1.8 \sin\alpha$, this value of x being true also for the case of calculating average concentrations for homogeneous films of finite thickness, that is also for the instances

(IV) $n = n_0\ (z \leqslant s\lambda), \quad n = 0\ (z > s\lambda);$

(V) $n = 0\ (z \leqslant s\lambda), \quad n = n_0\ (z > s\lambda).$

Table 2.7
Comparison of values $\bar{n}$ and f (in n_0 units) with $x = 1.79 \sin\alpha$

Case	f / n	α^0	S 0.125	0.25	0.50	1.0	2.0	4.0
I[a]	f	25	0.13	0.24	0.41	0.62	0.79	0.89
	n		0.08	0.16	0.33	0.62	0.81	0.91
	f	45	0.09	0.16	0.28	0.47	0.67	0.82
	n		0.05	0.10	0.20	0.40	0.68	0.84
	f	65	0.07	0.13	0.23	0.39	0.60	0.78
	n		0.04	0.08	0.15	0.31	0.59	0.80
II	f	25	0.87	0.76	0.59	0.38	0.21	0.11
	n		0.92	0.84	0.67	0.38	0.19	0.09
	f	45	0.91	0.84	0.72	0.53	0.33	0.18
	n		0.95	0.90	0.80	0.60	0.29	0.16
	f	65	0.93	0.87	0.77	0.61	0.40	0.22
	n		0.96	0.92	0.85	0.69	0.41	0.20
III	f	25	0.23	0.37	0.54	0.70	0.83	0.90
	n		0.16	0.31	0.52	0.70	0.83	0.91
	f	45	0.15	0.25	0.43	0.59	0.74	0.85
	n		0.10	0.20	0.36	0.57	0.74	0.86
	f	65	0.12	0.22	0.36	0.52	0.69	0.82
	n		0.08	0.15	0.30	0.50	0.69	0.89

[a] Accepted $n_1 = 0$ [see (2.53) and (2.54)].

Table 2.7 gives a comparison of values $\bar{n}$ and f for some cases. The value $x \sim 1.8 \sin\alpha$ minimizes the absolute error $(f - \bar{n}_{av})$. If it is necessary to minimize the relative error $|\bar{n}/f|$, then value $\bar{n}$ is determined for a layer of $y\lambda$ thickness.

An analysis shows that it is impossible to find one value y that would be common for various types of $n = \varphi\ (z, s, n)$. In particular, let us study the most important instance, when a rapid decrease of concentration of the studied element is observed along with an increase in z. This agrees with instances (I) and (III) with values of $s \leqslant 0.5$.

Functions $n = \varphi\ (z, s, n)$ of this type can be singled out experimentally on the basis of an essential lessening of the intensity ratio with an increase of angle α: for the instances studied by us, intensity decreases by more than 1.7 times. In these cases it is expedient to use the value y determined for a layer with a thickness $\lambda y = 1.2\lambda \sin\alpha$. In the event of the intensity dropping less than 1.7 times when α is increasing from 25 to 65°, it is expedient to leave the evaluation of $\lambda y = 1.8\lambda \sin\alpha$. In those cases when an increase of intensities is observed* [see instances (V) and (II)] with an increasing α in the event of a noticeable growth (for instance, by more than 70%) during transition from 25 to 65°, it is expedient to use a layer $\lambda y = 2\lambda \sin\alpha$ thick to evaluate $\bar{n}$. If intensity increases in a lesser degree, it is expedient to take the evaluation $\lambda y = 1.8\lambda \sin\alpha$. Use of the above-stated

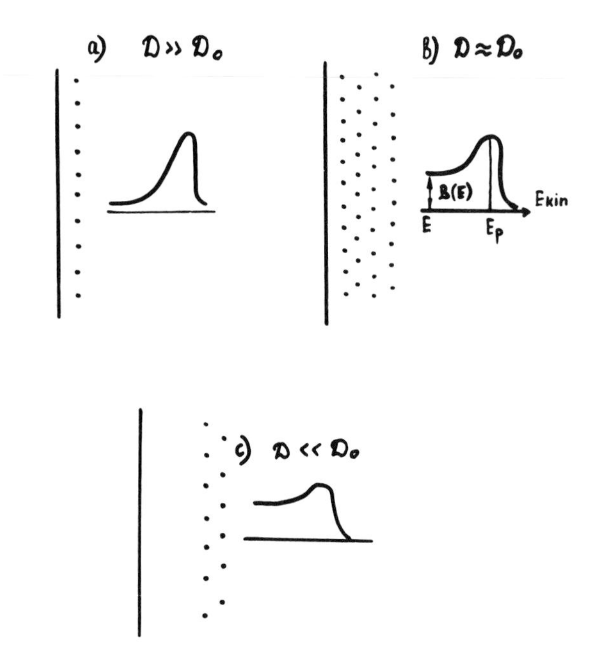

Figure 2.12. Dependence of the form of the X-ray photoelectron line of impurity atoms x (designated by dots) on their in-depth distribution. The vertical line complies with the position of the surface.

---

*It is expedient to measure angular dependence of the line intensity of the element under investigation with respect to the line of an element evenly distributed in the surface layer. This diminishes errors connected with the roughness of the sample, its uneven irradiation, etc.

assessments of the value y makes it possible to determine the ratio $\bar{n}_i/f_i$ with a precision of not less than 20% [83].

Reference [86] proposes a method of obtaining information on the distribution of elements throughout the depth of the sample from the distribution of photoelectrons at kinetic energies near the main maximum. This method is based on results obtained by the authors earlier [87] when analysing the energy distributions of photoelectrons that have left the sample. The inelastic and elastic collisions are considered. It was shown that the photoelectron distribution greatly depends on the path traversed by the photoelectron in the sample before reaching the surface.

Figure (2.12) schematically presents the X-ray photoelectron spectrum in the region of the main peak with $E_k = E_p$, where E_k is the photoelectron's kinetic energy. To characterize the properties of the homogeneous medium through which the photoelectron travels it is suggested to use the parameter

$$D(\Delta E) = A_p/B(E) \tag{2.60}$$

where A_p is the area of the main line while $B(E)$ is the intensity of the photoelectron flux at the distance $\Delta E = E_p - E$ from the main line (see Fig. 2.12).

Cited are measures of the value $D(\Delta E = 30\,\text{eV})$ for the polycrystalline pure metals Ni, Cu, Mo, Ag, Ta, W, and Pt, using the excitation radiation Mg Kα. It was found that for all metals (18 lines with E_p ranging from 300 to 1200 eV) the value $D(E = 30\,\text{eV})$ amounted to $D_0 = 23.0\,\text{eV}$ with a precision of $\approx 25\%$. The discovery of the existence of a 'universal' value of D_0 made it posible to propose a method of qualitatively determining the distribution of X atoms in the matrix Y on the basis of comparison of values $D(\Delta E = 30\,\text{eV})$ for X lines with the value D_0. Some possible profiles of the concentration of X elements throughout the depth of the sample and the expected ratios between D and D_0 for these profiles are presented in Fig. (2.12). Measurements of parameters D for samples having the following laminated structure were made to illustrate the suggested method: Al sublayer – Ag monolayer – Al film of thickness d from 0 to 7 nm. As predicted, with $d = 0$ the measured value $D = A/B$ turned out to be equal to $50\,\text{eV} > D_0$, this agreeing with case (a) in Fig. (2.12). With the growth of d the parameters D measured for the Ag 3d line drop sharply to 15–10 eV, thus agreeing with case (c) in Fig. (2.12), $D < D_0$.

Reference [88] presents the most general approach to the quantitative restoration of element concentration in depth profiles on the basis of X-ray photoelectron spectroscopy data. The flat sample of an AB binary alloy is divided into parallel surface layers of thickness x_0 and the angular distributions of the intensity ratios of lines I_A and I_B can be presented as:

$$I_A(\alpha_i)/I_B(\alpha_i) = \sigma_A/\sigma_B \sum_{j=1}^{n} \{N_A(j)[1 - \exp(-x_0/\lambda)\sin\alpha_i]$$
$$\prod_{k=1, j\geq 2}^{j=1} \exp\left(-\frac{x_0}{\lambda \sin\alpha_i}\right)\Bigg\} \Bigg/ \sum_{j=1}^{n} \left\{N_\theta(j)\left[1 - \exp\left(-\frac{x_0}{\lambda\sin\alpha_i}\right)\right]\cdot\right.$$
$$\left.\prod_{k=1, j\geq 2}^{j=1} \exp\left(-\frac{x_0}{\lambda\sin\alpha_i}\right)\right\} \tag{2.61}$$

where α_i is the exit angle of the photoelectrons, σ_A and σ_B are photoionization cross-section, and n is the number of layers considered. By minimizing the functional we can find the values of the concentrations of the elements $N_A(j)$ and $N_B(j)$ in layer j:

$$F = \sum_i [(I_A(\alpha_i)/I_B(\alpha_i))_{exper} - I_A(\alpha_i)/I_B(\alpha_i)]^2 \qquad (2.62)$$

where the sum includes all experimental points. But it was established in [86] that the direct use of the expressions (2.61) and (2.62) in the minimization procedure, for instance by the simplex method, results in difficulties caused by instability with respect to input data. Thus, in the event of a change in the ratio $(I_A(\alpha_i)/I_B(\alpha_i))_{exp}$ by a mere 0.001%, this being so infinitely small compared with typical experimental errors as to be ignored, absolutely different in-depth profiles of elements A and B can be obtained.

The only method of overcoming this instability was found in using some *a priori* information on the dependence of concentrations N_A and N_B on depth x during minimization. It has been found that at least two restrictions should be

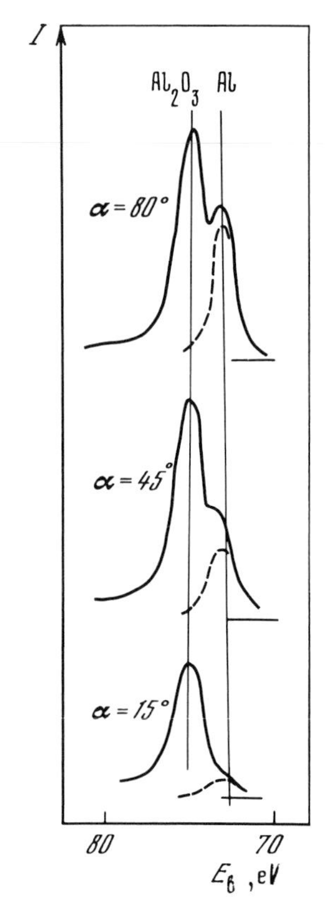

Figure 2.13. Dependence of Al 2p line intensity in oxide and metal for metallic aluminum, covered with a film of oxide, on angle α.

imposed on the functions $N_A(x)$ and $N_B(x)$. First, they should be limited ($0 \leqslant N_A, N_B \leqslant \text{const}$) and, second, monotonously increasing or monotonously decreasing, that is they should not have local extremums near the sample's surface.

The possibilities of this evolved method of restoring profiles proceeding from the angular distributions of intensities of X-ray photoelectron spectra are illustrated by the example of the systems Ag – Al_2O_3, SiO_2 – Si, and Cu – Ni. It is shown that an in-depth resolution of the order of $\lambda/3$ is thus obtained, λ being the length of the mean free path of photoelectrons without inelastic collisions. It appears promising to apply this method more broadly to more general distributions of concentrations (including also those with an extremum near the sample's surface) and to the instance of the dependence of λ on composition (here the problem of restoring the profile becomes nonlinear).

So X-ray photoelectron spectra carry information on the mean number of atoms in unit of volume of the surface layer, which has a thickness of the order of $1.8\lambda \sin \alpha$. The thickness of the measured surface layer, as has been noted already above, can be decreased both at the expense of α and at the expense of λ: the smaller α and E_{kin} ($E_{kin} > 150\,\text{eV}$), the higher the method's surface sensitivity. Figure (2.13) shows the simplest example, the growth of the relative intensity of the Al 2p line in an oxide as compared with the Al 2p line of metal in a sample of metallic aluminum covered with film Al_2O_3, and its dependence on angle α. As a somewhat more complex example let us consider [89] the results of studying the surface film on the intermetallid Ni_5Zr formed when the sample was heated in air for 320 minutes. An XPS analysis with subsequent ion sputtering (see Chapter 3) showed that the atomic ratio Ni/Zr is characterized by the existence of a minimum that is achieved after 1.5–2 min of sputtering [this corresponds roughly to 3–4 nm from a surface—Fig. (2.14)]. Figure (2.15) shows the dependence of the atomic ratio Ni/Zr on angle α and the duration of ion sputtering. The ratio Ni/Zr was determined by the intensity of lines Ni 2p/Zr 3d (A) and Ni 3p/Zr 3d (B). The values λ, $\lambda \sin 65°$, $\lambda \sin 25°$ are equal to 3.3, 3.0, and 1.3 nm for photoelectrons of the Ni 3p line and 1.9, 1.8, and 0.8 nm for photoelectrons of the Ni 2p line [E_{kin}(Ni 2p) $\ll$ E_{kin}(Ni 3p)]. In accordance with this the atomic ratio Ni/Zr in the surface layer before sputtering turned out to be higher in case A than in case B because the surface layer has a sharp gradient of concentrations Ni/Zr (Fig. 2.14).

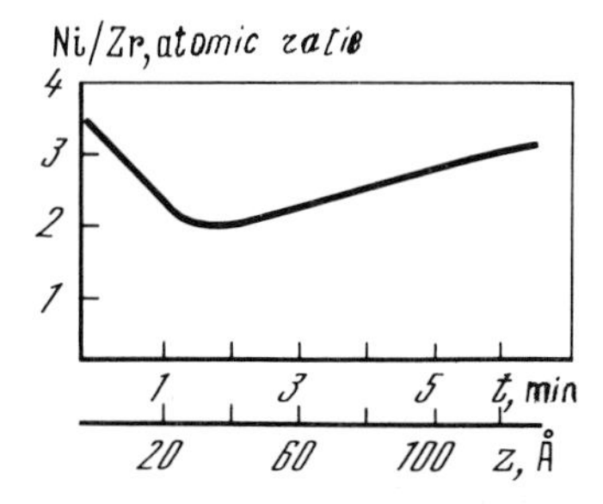

Figure 2.14. Dependence of Ni/Zr ratio on sputtering time.

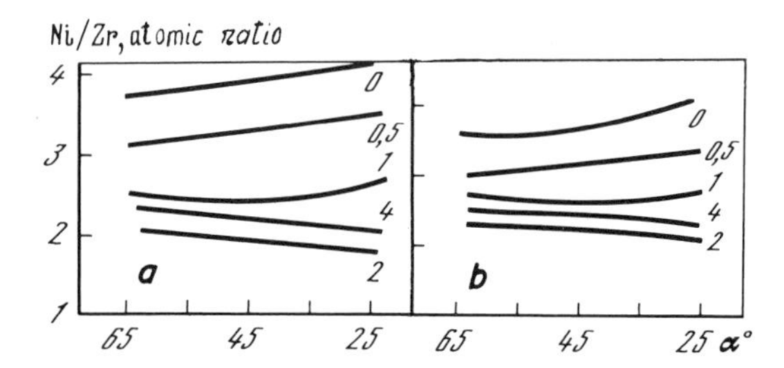

Figure 2.15. Dependence of Ni/Zr ratio on sputtering time and angle α. (a) Ni/Zr ratio determined by intensities Ni 2p/Zr 3d. (b) Ni 3p/Zr 3d ratio used. Figures by curves indicate sputtering time in minutes.

The presence of this gradient results in a growth of the ratio Ni/Zr with a decrease in α in both cases A and B. This is true also for measurements after 0.5 min of sputtering, but after 1 min the growth of Ni/Zr is observed only for small values of α. Moreover, it is somewhat better expressed for case B. The explanation is that in case B the value $1.8\lambda \sin\alpha$ at $\alpha = 65°$ is bigger than the distance (1.0–2.0 nm) from the new surface to the Ni/Zr ratio minimum (Fig. 2.14), but is smaller or roughly equal at $\alpha = 25°$, and for this reason the growth of ratio Ni/Zr begins at a certain interim angle (see Fig. 2.13). In case A angular dependence is extremely weak because the range $1.8\lambda \sin\alpha$ $(25 < \alpha < 65°)$ corresponds approximately to the position of Ni/Zr minimum. After 2 min of sputtering the sample's new surface approximately corresponds to the position of the Ni/Zr minimum and for this reason the value Ni/Zr decreases with the decrease in α. This is also true after 4 min of sputtering.

In conclusion let us note one useful way of drastically increasing the method's sensitivity, especially when studying small concentrations of metal atoms in solutions [80]. Glass plates with fixed chelates on the surface, that form complexes with the metal in the solutions, are made for this purpose. Since the metal's atoms are fixed on the surface, X-ray photoelectron spectra make it possible to determine concentrations in the range 10^{-6} in the initial solution. Such determinations have been demonstrated for bismuth, lead, tellurium, and mercury.

References

1. Fadley, C. S., Baird, R., Sichaus, W. *et al.* (1974) Surface analysis and angular distributions in X-ray photoelectron spectroscopy. *J. Electron Spectrosc. Rel. Phen.* **4**, 93–137.
2. Henke, B. L. (1972) Ultrasoft X-ray reflection, refraction and production of photoelectrons (100–1000 eV region). *Phys. Rev.* **A6**, 94–101.
3. Fadley, C. S. (1974) Instrumentation for surface studies. XPS angular distributions. *J. Electron Spectrosc. Rel. Phen.* **5**, 725–736.
4. Baird, R. J., Fadley, C. S., Kawamoto, S., and Mehta, M. (1975) Direct observation of surface profile effects on X-ray photoelectron angular distributions. *Chem. Phys. Lett.* **34**, 49–54.

5. Mehta, M. and Fadley, C. S. (1975) Enhancement of surface atom intensities in X-ray photoelectron spectra at low X-ray incidence angles. *Phys. Lett.* **A55**, 59–61.
6. Briggs, D. and Seah, M. P., Eds (1983) *Practical Surface analysis by Auger and XPS.* Wiley, Heyden.
7. Seah, M. P. (1980) The quantitative analysis of surfaces by XPS. *Surf. Interface Anal.* **2**, 222–239.
8. Nefedov, V. I. (1982) The physics of X-ray photoelectron analysis of surface composition. *Poverkhnost* **1**, 4–21 (in Russian).
9. Clark, D. T. and Shuttleworth, D. (1979) Angular and spatial-dependent intensity behavior for a prototype laterally inhomogeneous sample studied with an Es 200B spectrometer. *J. Electron Spectrosc. Rel. Phen.* **17**, 15–24.
10. Baird, R. J. and Fadley, C. S. (1977) X-ray photoelectron angular distributions with dispersion-compensating X-ray and electron optics. *J. Electron Spectrosc. Rel. Phen.* **1**, 39–66.
11. Yarzhemsky, V. G. (1980) Calculation of photoionization cross-sections in chemical compounds. Synopsis of thesis for the degree of candidate of sciences (chemistry). Institute of General and Inorganic Chemistry, Moscow (in Russian).
12. Pindzola, M. S. and Kelly, H. P. (1975) Comparison of the acceleration, length and velocity forms of the dipole operations in photoionization calculations. *Phys. Rev.* **A12**, 1414–1424.
13. Mehta, M., Fadley, C. S., and Bagus, P. S. (1976) Photoelectron peak intensities and atom–ion overlaps: analysis of various approximations. *Chem. Phys. Lett.* **37**, 21–28.
14. Fadley, C. S. (1974) Photoelectric cross sections and multielectron transitions in sudden approximation. *Chem. Phys. Lett.* **25**, 225–230.
15. Manson, S. T. (1976) Satellite lines in photoelectron spectra. *J. Electron Spectrosc. Rel. Phen.* **9**, 21–28.
16. Nefedov, V. I., Sergushin, N. P., Band, I. M., and Trzhaskovskaya, M. B. (1973) Relative intensities in X-ray photoelectron spectra. *J. Electron Spectrosc. Rel. Rhen.* **2**, 383–403.
17. Nefedov, V. I., Sergushin, N. P., Salyn, Ya. V. *et al.* (1975) Relative intensities in X-ray photoelectron spectra. Pt II. *J. Electron Spectrosc. Rel. Phen.* **7**, 175–185.
18. Scofield, J. H. (1976) Hartree–Slater subshell photoionization cross sections at 1254 and 1487 eV. *J. Electron Spectrosc. Rel. Phen.* **8**, 129–137.
19. Band, I. M., Kharitonov, Yu. I., and Trzhaskovskaya, M. B. (1979) Photoionization cross sections and photoelectron angular distributions for X-ray line energies in the range 0.132–4.509 eV. Targets: $1 \leqslant z \leqslant 100$. *At. Data Nucl. Data Tables* **23**, 443–503.
20. Nefedov, V. I. and Yarzhemsky, V. G. (1977) Relative intensities in X-ray photoelectron spectra. Pt III. *J. Electron Spectrosc. Rel. Phen.* **11**, 1–11.
21. Nefedov, V. I. and Yarzhemsky, V. G. (1977) Theoretical calculations of relative intensities in ESCA. *Phys. Scr.* **16**, 291–295.
22. Yarzhemsky, V. G. and Nefedov, V. I. (1978) Atomic shell photoionization cross-sections for FK_a line. *Zh. Strukt. Khim.* **15**, 934–937 (in Russian).
23. Yarzhemsky, V. G., Nefedov, V. I., Band, I. M., and Trzhaskovskaya, M. B. (1980) Relative intensities in X-ray photoelectron spectra. Pt V. *J. Electron Spectrosc. Rel. Phen.* **18**, p. 173–177.
24. Sergushin, N. P. and Nefedov, V. I. (1976) X-ray photoelectron quantitative analysis of the surface of solids. *Zh. Analit. Khim.* **31**, 2198–2204 (in Russian).
25. Evans, S., Pritchard, R. G., and Thomas, J. M. (1978) Relative differential subshell photoionization cross-sections (MgK_α) from lithium to uranium. *J. Electron Spectrosc. Rel. Phen.* **14**, 341–358.

26. Wagner, C. D. (1977) Factors affecting quantitative determinations by X-ray photoelectron spectroscopy. *Anal. Chem.* **49**, 1282–1290.
27. Elliot, I., Doyle, C., and Andrade, J. D. (1983) Calculated core-level sensitivity factors for quantitative XPS using an HP 5950B spectrometer. *J. Electron Spectrosc. Rel. Phen.* **28**, 303–316.
28. Wagner, C. D. (1983) Sensitivity factors for XPS—analysis of surface atoms. *J. Electron Spectrosc. Rel. Phen.* **32**, 99–102.
29. Wagner, C. D., Davis, L. E., Zeller, M. V. *et al.* (1981) Ampirical atomic sensitivity factors for quantitative analysis by ESCA. *Surf. Interface Anal.* **3**, 211.
30. Brunix, E. and van Eenbergen, A. (1983) Relative intra element cross-sections and binding energies obtained from XPS wide scan spectra. *Spectrochim. Acta*, **38B**, 821–830.
31. Reilman, R. F., Masezone, A., and Manson, S. T. (1976) Relative intensities in photoelectron spectroscopy of atoms and molecules. *J. Electron Spectrosc. Rel. Phen.* **9**, 389–394.
32. Stanley, W. R. and Shirley, D. A. (1977) Comparison of final state approximations in the calculation of total and differential photoemission cross sections of neon. *J. Chem. Phys.* **66**, 2378–2386.
33. Yarzhemsky, V. G., Nefedov, V. I., Amusia, M. Ya. *et al.* (1980) Relative intensities in X-ray photoelectron spectra. Pt VI. *J. Electron Spectrosc., Rel. Phen.* **19**, 123–254.
34. Ebel, M. F. (1978) Zur Bestimmung der reduzierten dicke D/λ dunner Schichten mittels XPS. *J. Electron Spectrosc. Rel. Phen.* **14**, 287–322.
35. Ebel, M. F. and Wernisch, J. (1981) Shading at different take-off angles in XPS. *Surf. Interface Anal.* **3**, 191–193.
36. Bernardez, L. S., Ferron, Y., Goldberg, E. S. *et al.* (1984) The effect of surface roughness on XPS and AES. *Surf. Sci.* **139**, 541–548.
37. Seah, M. P. and Dench, W. A. (1979) Quantitative electron spectroscopy of surfaces. *Surf. Interface Anal.* **1**, 2–11.
38. Baschenko, O. A. and Nefedov, V. I. (1979) Relative intensities in X-ray photoelectron spectra. Pt IV. *J. Electron Spectrosc. Rel. Phen.* **17**, 405–420.
39. Baschenko, O. A. and Nefedov, V. I. (1980) Relative intensities in X-ray photoelectron spectra. Pt VII. *J. Electron Spectrosc. Rel. Phen.* **21**, 153–169.
40. Baschenko, O. A. and Nefedov, V. I. (1982) Relative intensities in X-ray photoelectron spectra. *J. Electron Spectrosc. Rel. Phen.* **27**, 109–118.
41. Hill, J. M., Royce, D. G., Fadley, C. S. *et al.* (1976) Properties of oxidized silicon as determined by angular dependent X-ray photoelectron spectroscopy. *Chem. Phys. Lett.*, **44**, 225–231.
42. Powell, C. J. (1974) Attenuation lengths of low-energy electrons in solids. *Surf. Sci.* **44**, 29–46.
43. Londry, F. and Slavin, A. J. (1983) The determination of electron inelastic mean free path using evaporated films. *J. Vac. Sci. Technol.* **A1**, 44–48.
44. Pate, B. B., Oshima, M., Silberman, J. A. *et al.* (1984) Carbon 1s studies of diamond (111). *J. Vac. Sci. Technol.* **A2**, 957–960.
45. Steinhardt, R. G., Hudis, J., and Perlman, M. L. (1972) Attenuation of low energy electrons by solids. *Phys. Rev.* **B5**, 1016–1021.
46. Evans, S., Pritchard, R. G., and Thomas, J. M. (1977) Escape depths of X-ray photoelectrons and relative photoionization cross-sections for 3p-subshell of elements of first long period. *J. Phys.* **C10**, 2483–2498.
47. Thorn, R. J. (1983) Determination of ratios of photoelectron mean free paths from orbital ionization cross-sections. *J. Electron Spectrosc. Rel. Phen.* **31**, 207–220.

48. McGuire, E. J. (1971) Inelastic scatering of electrons and protons by elements He to Na. *Phys. Rev.* **A3**, 267–279.
49. McGuire, E. J. (1977) Electron ionization cross sections in Born approximation. *Phys. Rev.* **A16**, 62–72.
50. McGuire, E. J. (1979) Scaled electron ionization cross sections in the Born approximation for atoms with $55 \leqslant Z \leqslant 102$. *Phys. Rev.* **A20**, 445–456.
51. Lotz, W. (1970) Electron impact ionization cross sections for atoms up to $Z = 108$. *Z. Phys.* **232**, 101–113.
52. Penn, D. R. (1976) Electron mean free paths for free electron line materials. *Phys. Rev.* **B13**, 5248–5254.
53. Penn, D. R. (1976) Quantitative chemical analysis by ESCA. *J. Electron Spectrosc. Rel. Phen.* **9**, 29–40.
54. Ashley, J. C., Tung, C. J., and Ritchie, R. H. (1978) Electron inelastic mean free path and energy losses in solids. I. Aluminium metal. *Surf. Sci.* **81**, 409–428.
55. Tung, C. J., Ashley, J. C., and Ritchie, R. H. (1979) Electron inelastic mean free paths and energy losses in solids (II). Electron gas statistical model. *Surf. Sci.* **81**, 427–433.
56. Ashley, J. C. (1982) Simple model for electron inelastic mean free paths. *J. Electron Spectrosc. Rel. Phen.* **28**, 177–194.
57. Ashley, J. C. and Tung, C. J. (1982) Electron inelastic mean free paths in several solids for $200\,\text{eV} < E < 10\,\text{KeV}$. *Surf. Interface Anal.* **4**, 52–55.
58. Wagner, C. D., Davis, L. E., and Riggs, W. M. (1980) The energy dependence of the electron mean free path. *Surf. Interface Anal.* **2**, 53–55.
59. Quinn, J. J. (1962) Range of excited electrons in metals. *Phys. Rev.* **126**, 1453–1457.
60. Brundle, C. R., Hopster, H., and Swalen, J. D. (1979) Electron mean free path lengths through monolayers of cadmium arachidate. *J. Chem. Phys.* **70**, 5190–5196.
61. Cadman, P., Gossedge, G., and Scott, J. D. (1975) The determination of the photoelectron escape depths in polymers and other materials. *J. Electron Spectrosc. Rel. Phen.* **13**, 1–6.
62. Szajman, J., Jenkin, J. G., Liesegang, J., and Leckey, R. C. G. (1978) Electron mean free paths in Ge in the range 70–1400 eV. *J. Electron Spectrosc. Rel. Phen.* **14**, 41–48.
63. Vasquez, R. R. and Grunthaner, F. J. (1980) Intensity analysis of XPS spectra to determine oxide non-uniformity; application to SiO_2/Si interface. *Surf. Sci.* **99**, 681–688.
64. Roberts, R. F., Allara, D. L., Pryde, C. A. *et al.* (1980) Mean free path for inelastic scattering of 1.2 keV electrons in thin poly(methylmethacrylate) films. *Surf. Interface Anal.* **2**, 5–10.
65. Norman, D. and Woodruff, D. P. (1978) Energy dependence on electron inelastic scattering mean free path using synchrotron radiation photoelectron spectroscopy. *Surf. Sci.* **75**, 179–198.
66. Szajman, J., Jenkin, J. G., Leckey, R. C. G., and Liesegang, J. (1980) Subshell photoionization cross-sections, electron mean free paths and quantitative X-ray photoelectron spectroscopy. *J. Electron Spectrosc. Rel. Phen.* **19**, 3–93.
67. Szajman, J., Leckey, R. C. G., Liesegang, J., and Jenkin, J. G. (1980) Electron mean free path in CdTe 350–1450 eV. *J. Electron Spectrosc. Rel. Phen.* **20**, 323–326.
68. Fraser, W. A., Florio, J. V., Delgass, W. N., and Robertson, W. D. (1973) Surface sensitivity and angular dependence of X-ray photoelectron spectra. *Surf. Sci.* **36**, 661–674.
69. Brunner, J. and Zogg, H. (1974) Angular dependence of X-ray photoelectrons. *J. Electron Spectrosc. Rel. Phen.* **5**, 911–922.

70. Hollinger, G., Jugnet, Y., Petrosa, P. *et al.* (1977) X-ray photoelectron characterization of very thin silicon oxide films. *Analysis* **5**, 2–10.
71. Feibelman, P. J. (1973) Spatial variation of the electron mean free path near a surface. *Surf. Sci.* **36**, 558–568.
72. Ebel, H., Ebel, M. F., Wernisch, J., and Jablonski, A. (1984) The influence of elastic scattering of electrons on measures X-ray photoelectrons signals. *J. Electron Spectrosc. Rel. Phen.* **34**, 355–362.
73. Baschenko, O. A., Machavariani, G. V., and Nefedov, V. I. (1984) New technique for investigation of angular distribution of photoemission from solids. *J. Electron Spectrosc. Rel. Phen.* **34**, 305–308.
74. Dwyer, V. M. and Mathew, J. A. D. (1984) The effects of elastic block scattering on the auger or X-ray photoelectron spectra of solids. *Surf. Sci.* **143**, 57–83.
75. Ebel, H., Ebel, M. F., Wernisch, J., and Yablonski, A. (1984) Quantitative XPS analysis considering elastic scattering. *Surf. Interface Anal.* **6**, 140–143.
76. Baschenko, O. A. and Nefedov, V. I. (1982) Effective mean free paths of photoelectrons in flat films. *Poverkhnost* **1**, 87–93 (in Russian).
77. Leckey, R. C. G. (1976) Subshell photoionization cross-sections of elements for Al K_α radiation. *Phys. Rev.* **A13**, 1043–1051.
78. Powell, C. J. and Larson, P. E. (1978) Quantitative surface analysis by X-ray photoelectron spectroscopy. *Appl. Surf. Sci.* **1**, 186–201.
79. Minnhagen, P. (1976) Exact numerical solution of a Nizieres–De Dominicis-type model problem. *Phys. Lett.* **A56**, 327–329.
80. Hercules, D. M. (1974) Electron spectroscopy for chemical analysis. *J. Electron Spectrosc. Rel. Phen.* **5**, 811–826.
81. Powell, C. J., Erickson, N. E., and Madey, T. E. (1979) Results of joint Auger/ESCA round robin sponsored by ASTM committee E-42 on surface analysis. Pt I. ESCA results. *J. Electron Spectrosc. Rel. Phen.* **17**, 361–403.
82. Nefedov, V. I. (1982) A comparison of results of an ESCA study of nonconducting solids using spectrometers of different constructions. *J. Electron Spectrosc. Rel. Phen.* **25**, 29–49.
83. Nefedov, V. I. (1981) X-ray photoelectron analysis of surface layers with concentration gradient. *Surf. Interface Anal.* **3**, 72–75.
84. Finster, J., Lorenz, P., and Meisel, A. (1979) Quantitative ESCA surface analysis applied to catalysts investigation of concentration gradients. *Surf. Interface Anal.* **1**, 179–184.
85. Nefedov, V. I. (1982) X-ray photoelectron analysis of surface with a concentrations gradient. *Poverkhnost* **1**, 115–121 (in Russian).
86. Tougaard, S. and Ignatiev, A. (1983) Concentration depth profiles by XPS. A new approach. *Surf. Sci.* **129**, 355–365.
87. Tougaard, S. and Ignatiev, A. (1983) Background intensities in XPS spectra from homogeneous metals. *Surf. Sci.* **124**, 451–460.
88. Pijolat, M. and Hollinger, G. (1981) New depth-profiling method by angular-dependent X-ray photoelectron spectroscopy. *Surf. Sci.* **105**, 114–128.
89. Nefedov, V. I., Lunin, V. V., and Chulkov, N. G. (1980) Dependence of surface composition of Zr–Ni intermetallic compounds. *Surf. Interface Anal.* **2**, 207–211.

Chapter 3

Ion sputtering and in-depth analysis

X-ray photoelectron spectra, as has been noted in the previous chapter, carry information about the surface layer to a depth of about $2\lambda \sin \alpha$, where λ is the mean free path of the electron and α the photoelectron exit angle. Since λ is only about 1.0–4.0 nm, it is necessary to remove the surface layer in order to analyse deeper layers. Different methods can be used for the purpose [1], but ion sputtering is the most universal. Figure (3.1) shows changes in the intensity of C 1s, O 1s, Si 2p, and Al 2p X-ray photoelectron lines with time T being the time of sputtering of the specimen with Ar^+ ions. The specimen consists of an overlayer of polymer film containing Si, C, and O, deposited on aluminium. Since the film thickness is several tens of nm, the Al 2p line at $T = 0$ is absent. The initial change in intensity is caused by the removal of a contamination layer and the plateau represents the polymer film, which is characterized by constant C 1s, O 1s, and Si 2p intensities. When the bulk of the polymer film has been sputtered, the intensities of the C 1s and Si 2p lines decrease and the Al 2p line appears. The O 1s line grows more intense because metallic aluminum has an Al_2O_3 overlayer. The intensity of the O 1s line drops after the layer of Al_2O_3 has been removed, while the Al 2p intensity continues to grow. Obviously, information on the distribution of elements in the sample can be obtained by analysing the dependence of line intensity on sputtering time.

The interaction of ions with solids is considered in a large number of papers and reviews (see e.g., [2–8] and bibliography up to 1978 in [9]). This chapter will be restricted to the application of ion sputtering to XPS in-depth analysis.

Ion sputtering has a substantial role to play in obtaining clean metallic surfaces (see review [10], which lists appropriate methods for 74 elements).

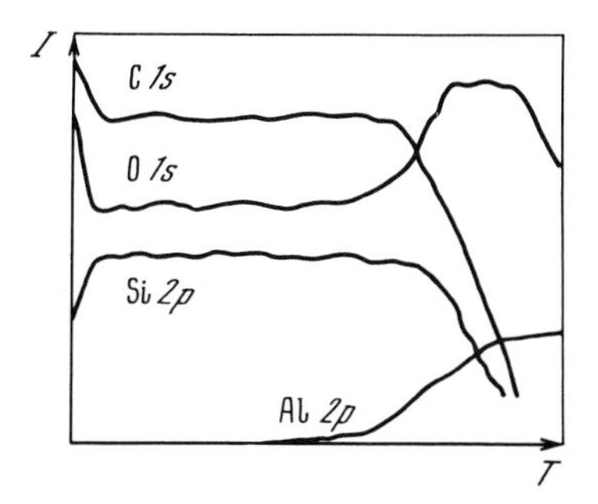

Figure 3.1. Dependence of X-ray photoelectron line intensities on sputtering time.

3.1. Factors influencing sputtering speed

As ions interact with solid surfaces, energy from incoming ions (usually several per cent) is imparted to the atoms of solids, and solids are sputtered. As a rule, free neutral particles and, to a lesser extent, positive and negative ions are formed. These particles can consist of one or more atoms. The sputtering speed V can be represented as [5]

$$V = 0.06\,\gamma\, IA/\rho \qquad A/\text{min} \tag{3.1}$$

where γ is the sputtering coefficient, atom/ion; I is the current of the primary ion beam in $A\,\text{cm}^{-2}$; A is the atomic or molecular mass, in $\text{g}\,\text{mol}^{-1}$; and ρ is the density of the target material is $\text{g}\,\text{cm}^{-3}$. V can be calculated experimentally from the decrease in target mass ΔG, or from the drop in the intensity of the XPS line in passage through the boundary of a film of known thickness t during time T:

$$V = \Delta G/\rho\, ST,\; V = t/T, \tag{3.2}$$

where S is the area of the sputtering spot.

A simple experimental determination of sputtering speed is also possible when the concentration gradient of one of the elements being analysed is known [11].

Since it is often hard to determine V experimentally, it often has to be evaluated on the basis of the published values of γ according to Equation (3.1), or on the basis of the equation

$$V_2 = V_1 \gamma_2/\gamma_1, \tag{3.3}$$

where V_1 is determined experimentally.

Let us examine the dependence of γ on various factors. The value γ usually [5] stands in proportion to $mM/(m + M)$, where M and m are the masses of the target atom and the sputtering ion, respectively. This value grows as m gets larger, which is why sputtering with heavy ions, other conditions being equal, is usually more efficient, especially with large values of M. Sputtering for purposes of XPS analysis is done with ions (usually of noble gases) having energies from several hundred eV to several keV. The value γ increases in this range as the energy E_{kin} of the primary ion beam increases, and the dependence of γ on E_{kin} is usually almost linear. A drop in γ with E_{kin} may be observed for some boundary E_{kin} values. The maximum value of E_{kin} is relatively low for the H^+ ion (2 keV) but far higher for ions of noble gases [6].

Table 3.1
Sputtering coefficients for Kr^+ (10 keV)

Oxide	γ(oxide)	γ(metal)	Oxide	γ(oxide)	γ(metal)
Al_2O_3	1.5	3.2	Ta_2O_5	2.5	1.6
MgO	1.8	8.1	TiO_2	1.6	2.1
MoO_3	9.6	2.8	UO_2	3.8	2.4
Nb_2O_5	3.4	1.8	V_2O_5	12.7	2.3
SiO_2	3.6	2.1	WO_3	9.2	2.6
SnO_2	15.3	6.5	ZrO_2	2.8	2.3

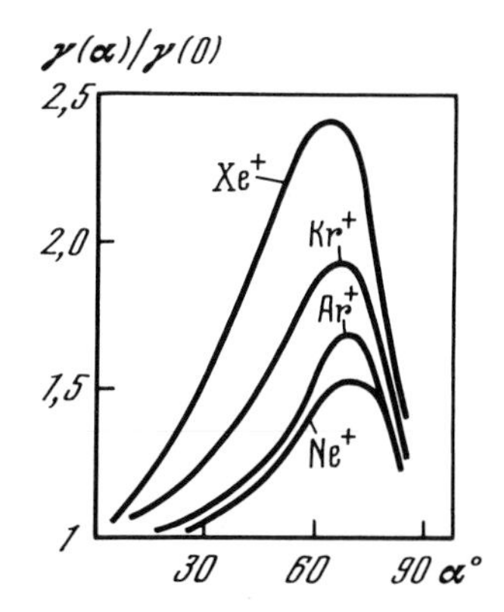

Figure 3.2. Dependence of the sputtering coefficient of ion-bombarded copper on angle α of ion fall. Angle α is measured normal to the surface.

The value γ for metals depends on the position of the metal in the Periodic Table. Metals of the 4th and 5th groups with maximum sublimation heat have minimum γ values. The γ values for oxides may be either higher or lower than the equivalent values for metals (Table 3.1 [12]).

Sputtering in monocrystals depends strongly on the crystal structure and lattice orientation. If the density of the lattice knots is high in the plane perpendicular to the ion beam, dispersion proceeds more efficiently; for instance, γ for Cu(111) is double that for Cu(110). Angular dependence also is observed in polycrystalline samples. According to [13] (Fig. 3.2), when copper is sputtered by ions of noble metals with $E_{\text{kin}} = 0.5 \div 2\,\text{keV}$, $\gamma(\alpha)$ reaches its maximum at $\alpha \sim 60 \div 70°$ (α is measured from the normal to the surface). It is necessary to stress that the value I in (3.1) is proportional to $\cos\alpha$ so that V grows in relation to $c\alpha = 0$ only at those values of α for which $\gamma(\alpha)/\gamma(0) > 1/\cos\alpha$.

It is necessary to avoid the oblique fall of the beam on the sputtered target because a sharp drop in the α value can be expected. Similar conclusions on the effect of angle on ion sputtering speeds were drawn in [14], which also studied the effect of specimen rotation. The temperature of the target usually has a slight effect on ion sputtering speeds. A review of thermal effects in ion sputtering is offered in [15].

At present there are a number of models (see bibliographies in [2, 6, 16]) explaining the dependence of γ on various factors, but the accuracy of computed γ values is small (collected experimental γ values are supplied in [2, 5, 6] and others). Tables (3.2) and (3.3) [5, 6] give γ values for metals at $E_{\text{kin}} = 1$ and $10\,\text{keV}$ respectively, in ions of noble gases. Values supplied by different authors are cited in some cases (Cu, Ni). This gives an idea of the accuracy of γ values.

To conduct an XPS analysis, it is necessary to ensure a smooth layer-by-layer sputtering of a part of the specimen that is larger than the spot from which the photoelectrons pass into the analyser. The area of the spot usually amounts to several mm^2. Hence it is necessary to carry out sputtering with an unfocused ion beam or with several ion sources simultaneously, and to use scanning, since the sputtering speed usually varies along the ion beam. The latter circumstance

Table 3.2
γ Coefficients at $E_{kin} = 1$ keV

Metal	He^+	Ne^+	Ar^+	Kr^+	Xe^+
Be	0.35	0.80	1.1	0.8	0.7
Al		1.13	1.90	1.53	
Si			1.0		
Ti			1.13		
Fe		0.84	1.34	1.44	
Ni	0.22	1.45	1.86	1.89	2.0
		1.24	2.18	1.73	2.22
			2.16		
Cu	0.65	2.75	3.64	3.62	3.42
		1.88	2.90	3.43	5.70
			3.2		
Ge			1.55		
Zr			1.06		
Nb			0.98		
Mo		0.43	1.14	1.41	1.63
Ag	1.8	2.4	3.8	4.7	
			4.7		
Cd			11.2		
Ta			0.91		
W			1.10		
Pt	0.08	0.85	2.0	2.35	2.52
Au		1.53	3.08	3.86	
			4.02		
Pb			4.2		

Table 3.3
Coefficients at $E_{kin} = 10$ keV

Metal	N_2	Ar	Kr	Xe
Cu	3.2	6.6	8.0	10.0
Ag		8.8	15.0	16.0
Au	3.7	8.4	15.0	20.0
Fe		1.0		
Mo		2.1		
Ti		2.1		

causes the appearance of craters. Although the analysis can proceed despite the presence of craters (see theoretical discussion in [17]), accuracy is badly affected. References [18, 19] describe methods that make it possible to ensure uniform sputtering. The following experiment is an effective test of the adequacy of the sputtering process.

It is necessary to make sure that the intensity of the substrate line after the sputtering of the surface film coincides with the intensity of the line in substrate without film. This test also guarantees the correctness of the adjustment of the sputtering spot. Sputtering is usually done with ions of a noble gas at a pressure of roughly 10^{-3} pa. Sputtering with plasma is sometimes used but it involves chemical reactions (reactive ion sputtering; see review in [20]).

3.2. Influence of ion bombardment on concentration profile of elements under examination

The initial distribution of concentration $n(z)$ of the elements under examination, depending on depth z in ion sputtering, can be altered as a result of the statistical character of sputtering in depth, atomic mix-up, and diffusion (see review in [21]). Let us first examine the statistical character of the sputtering effect.

Assume that only atoms in the surface monolayer are removed by ion bombardment. Obviously, after an atom from the first surface monolayer has been removed, an atom from the second monolayer can be ejected even though there still remain atoms from the first monolayer. As a result, the surface becomes quite rough (Fig. 3.3 [22]). The distribution of highs or lows z in relation to the average value $\bar{z} = an$ (where n is the number of monolayers and a the thickness of the monolayer), as references [19, 23] demonstrate, is governed by the Gauss law, $W(z)$ (Fig. 3.3), and the average statistical deviation 2σ for Δn equals

$$2\sigma = \Delta n = 2\sqrt{n}, \quad \Delta z = a\Delta n, \tag{3.4}$$

from which follows

$$\Delta z/\bar{z} = 2\sqrt{(a/\bar{z})}, \tag{3.5}$$

$$W(z) \sim \exp(-2z/\Delta z)^2. \tag{3.6}$$

If at $z = z_0$ there is a transition from one phase to another (Fig. 3.4) and if intensity I_A of the signal of atom A atom depends only on the number of A atoms

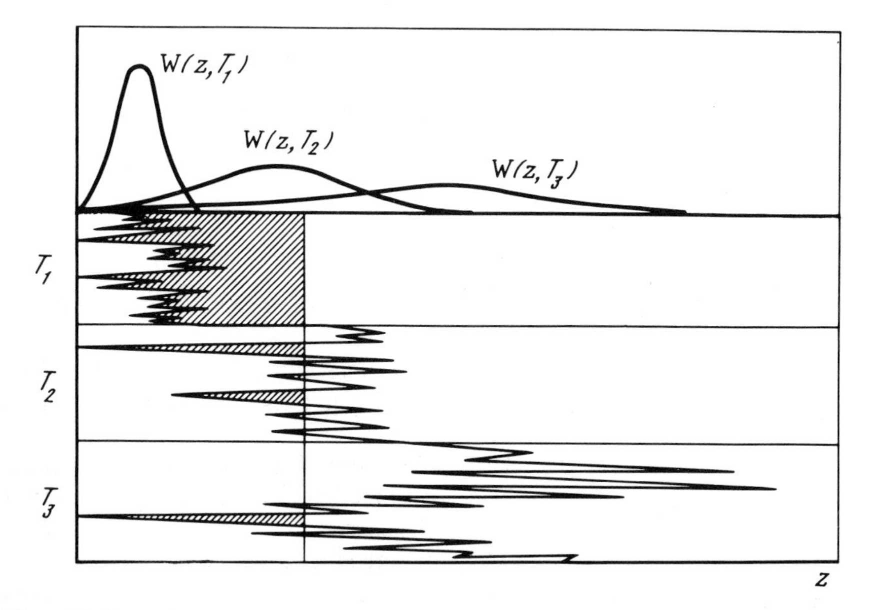

Figure 3.3. Dependence of distribution $W(z, T)$ on sputtering time $T_3 > T_2 > T_1$.

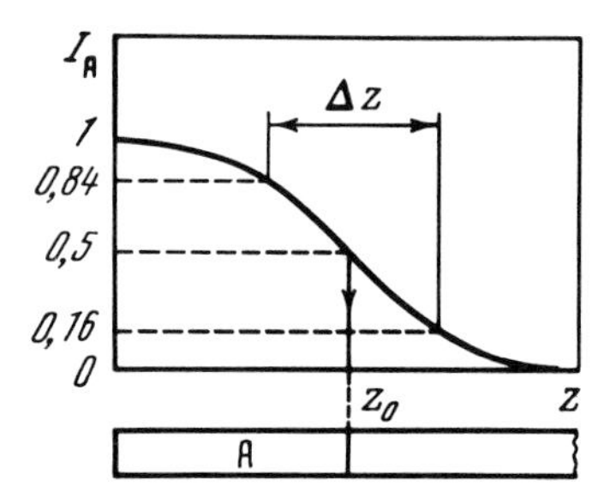

Figure 3.4. Determination of z_0 and Δz from the dependence of I_A on z.

in the first monolayer, then

$$I_A = \int_0^\infty W(z)\mathrm{d}z. \tag{3.7}$$

The dependence of I_A on z is shown in Fig. (3.4).

It is known from statistical theory that the area determined according to Fig. (3.4) contains 68% of the area of all the highs and lows, characterized by the value z, with respect to spot z_0 (see Fig. 3.3). An analysis of the function $I_A = f(z)$ thus makes it possible to determine $\Delta z/z$ experimentally (see Fig. 3.4). If the value z_0 is measured, it is possible to verify (3.5). Some experiments [19] accord with law (3.5), but reference [24] has established for V/Ti the following dependences: $\Delta z/z \sim z^{-1}$ at $z > 80$ nm and $\Delta z/z = \text{const}$ at $z < 80$ nm. In a study of Au films on Ni, reference [25] has also established a relationship between Δz on E_{kin} of bombarding ions: $\Delta z = a + b\ (E_{\text{kin}} z)^{1/2}$. Equation (3.5) will be more accurate if we bear in mind that not only ion bombardment but also sputtering itself is statistical. The equation then will be as follows [26]:

$$\Delta z/\bar{z} = 2\sqrt{a/\bar{z}}\sqrt{1+\gamma} \tag{3.8}$$

which takes account of the dependence of Δz on E_{kin} through sputtering coefficient γ but inflates Δz values.

Equations like (3.7) are not observed for intensities of XPS spectra* because they are sensitivie not only to the first monolayer. Proceeding from (2.33), the intensity I_T at time T after the beginning of sputtering of a film of initial thickness $t = z_0$ can be presented [22] as follows:

$$I_T = I_\infty[1 - \exp(\bar{V}T - t)/\lambda \sin\alpha] \tag{3.9}$$

where V is the average ion sputtering speed and I_∞ the intensity of the semi-infinite specimen.

Equation (3.9) does not take account of the statistical character of ion bombardment, i.e. the distribution of values $VT = z$ relative to the average value

*This equation holds only when $\lambda \approx a$, where λ is the electron mean free path and a the monolayer thickness (0.2–0.3 nm).

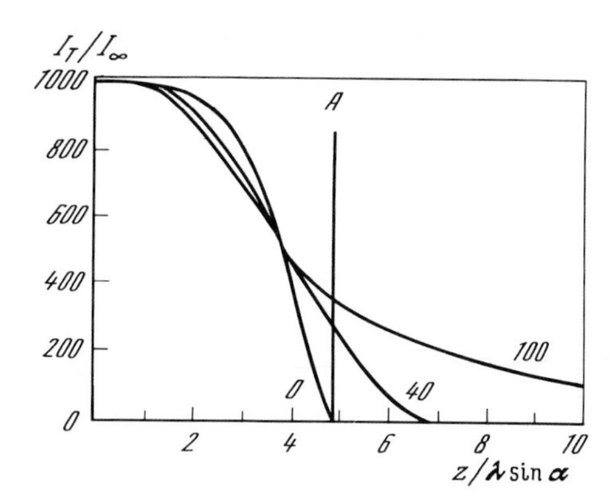

Figure 3.5. Dependence I_T/T on $z/\lambda \sin \alpha$ at different S values (figures next to curves). A = position of film interface.

$\bar{V}T$. I_T can be written correctly by integrating Equations (3.7) and (3.9):

$$I_T/I_\infty = \int [1 - \exp(z-t)/\lambda \sin \alpha] W(z,T)\mathrm{d}z \Big/ \int W(z,T)\mathrm{d}z,$$

$$W(z,T) = \exp[-2(z - \bar{V}T)/\Delta z]^2. \tag{3.10}$$

An equation of the type (3.10) was analysed in [22], which determined Δz as follows:

$$\Delta z = 0.01\, S\bar{V}T = 0.01\, S\bar{z},$$

where S characterizes surface roughness after ion bombardment. The integral in (3.10) was replaced with a summation in $z \pm \Delta z$.

The value $S = 100\%$ corresponds to the contribution of the surface layer itself to intensity I_T after any bombardment time T, though this contribution decreases along with T. A more even sputtering of the sample is observed with $S < 100\%$. Figure (3.5) shows the calculated dependence of I_T/I on z for $\lambda = 0.75$ nm and $\alpha = 42.3°$. The film boundary was determined at $z_0 = 2.5$ nm. The figure shows that the position of the boundary does not correspond to the value of sputtering time T, at which $(I_T/I_\infty)_T = \frac{1}{2}(I_T/I_\infty)_{max}$, though many experimental projects determine the boundary of the film in precisely this way.

If the substrate is an insulator and the film a conductor, as the film–substrate boundary is passed in sputtering there should appear on the substrate isolated spots of film due to the statistical character of sputtering and substrate roughness. Charging effects for these spots and the film itself are different, so signals from atoms of the film can be duplicated. The phenomenon was actually recorded in the observation of films of FeNi alloy on SiO_2 [27]. This phenomenon can also be used to determine the film–substrate boundary.

The value Δz and the entire formally recorded distribution profile of the concentration of the analysed element $n(z)$ depend on the ratio of the sputtering coefficients γ_1/γ_2. This problem is discussed with respect to alloys in [28].

The above method of calculating Δz takes no account of changes that occur in the initial distribution $n(z)$ of the atoms as thin the sample as a consequence of ion bombardment; the method is therefore rather formal. With an E_{kin} value of several hundred eV, these changes affect several dozen monolayers of the specimen. A major shortcoming is the 'knock-in' of atoms into the bulk of the sample, which causes 'tails' in the analysed distribution patterns. Examples of this phenomenon are cited e.g., in [5]. The 'knock-in' effect grows as E_{kin} increases. Reference [29], describes an analysis of the distribution of B atoms in Si by mass spectrometry of secondary ions, using ion sputtering with Ar^+. This study shows that a drop in the B signal to 0.01% of its maximum level is achieved with $z = 160$, 180, 220, and 270 nm for energies $E_{kin} = 5$, 10, 30, and 50 keV, respectively.

Changes in the distribution of $n(z)$ constitute a complex function of the character of the target, the atom being analysed, the ion used for sputtering, E_{kin}, γ, and the angle θ between the normal to the sample surface and the ion beam. To take account of all these factors, the sputtering process is modelled by the Monte Carlo method (see bibliography in [30]) and it is assumed that sputtering proceeds layer by layer with a constant speed and brings about a constant Δz for any given value of E_{kin}.

Let us look at the results of the calculations in [30]. According to data obtained from the perpendicular fall of an Ar^+ beam ($E_{kin} = 5$ keV) on Si, the most substantial changes for $n(z)$ monolayers of B atom in a matrix of Si at a distance z_0 from the surface

$$n(z) = \text{const at } z = z_0, n(z) = 0, z \neq z_0$$

are observed at $z_0 \approx 20\,n$ (where n is the number of monolayers). This roughly corresponds to the distance over which Ar^+ ions lose maximum energy. When $\theta = 60°$, the most significant changes in $n(z)$ occur at $z_0 = 0$. Figure (3.6)[30] shows the distribution of atom B ($z_0 = 50$) at different sputtering

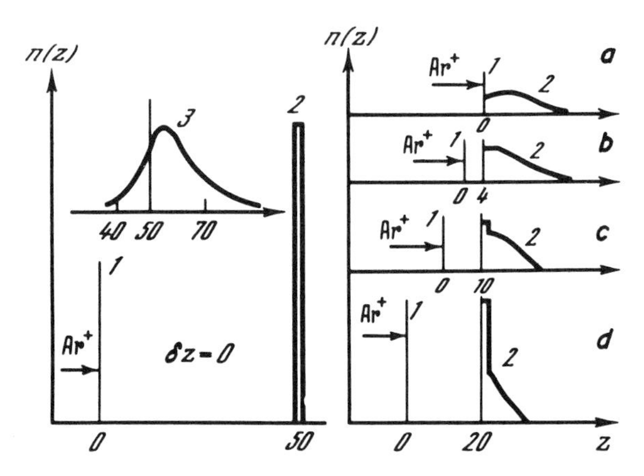

Figure 3.6. Monte Carlo calculation of distributions $n(z)$ of B atom in Si depending on the size of the sputtered layer. $\delta z = 1$ (a) 50; (b) 46; (c) 40; and (d) 30.
1 = surface; 2 = distribution $n(z)$; 3 = fixed distribution $n(z)$ by the secondary ion mass spectrometry method; z = depth in monolayers.

depths δz, and also the profile $n(z)$ expected e.g., for the secondary ion mass spectrometry method, which reflects the surface composition in the course of sputtering. The chart shows the formation of a 'tail' and also the shift of the distribution maximum into the bulk of the sample. Calculations show that the shift reaches a maximum when the ion beam falls on the sample at the right angle. The rise of E_{kin} has little effect on this shift but increases the blurring of $n(z)$ both for a distribution of the Equation (3.1) type and for the case of film sputtered on substrate [31] (this increases the half-width of distribution $n(z)$). The latter effect is also observed with a decrease in γ, all the other conditions being unchanged, because sputtering time is extended as a result. These findings agree with experimental data.

The 'knock-in' effect strongly depends on the matrix. Reference [32] shows that the sputtering away of 12 monolayers of Mo on W, Cu, and Al proceeds in different ways. Especially long 'tails' were recorded with Mo on Al.

Better resolution can be achieved in a number of cases by decreasing Δz and the 'knock-in' effect through sputtering with reactive ions, which react with the surface. In particular, reference [33] demonstrates that with $E_{kin} = 1 \div 2\,\text{keV}$, sputtering with N_2^+ ions ensures a far better resolution than that obtained with Ar^+ ions. Figure (3.7) [33] illustrates the results of the sputtering of a specimen composed of a set of Ge, and Nb layers, each with a thickness of 10 nm. The presence of 'tails' in the case of sputtering with Ar^+ makes it impossible to detect the layered structure of the specimen from Nb spectra.

The dependence of signal intensity on sputtering depth is also a function of the specimen shape. If small spherical grains covered with film are exposed to sputtering, the dependence of the signal intensity from atoms in the bulk of the grains and from the film surface during sputtering has the form shown in Fig. (3.8) [34]. It is important to note that the signal from the surface film is weakened very little with increase in sputtering time. This result was corroborated by an experiment in which TiO_2 pellets covered with an SiO_2 or Al_2O_3 film were sputtered [34].

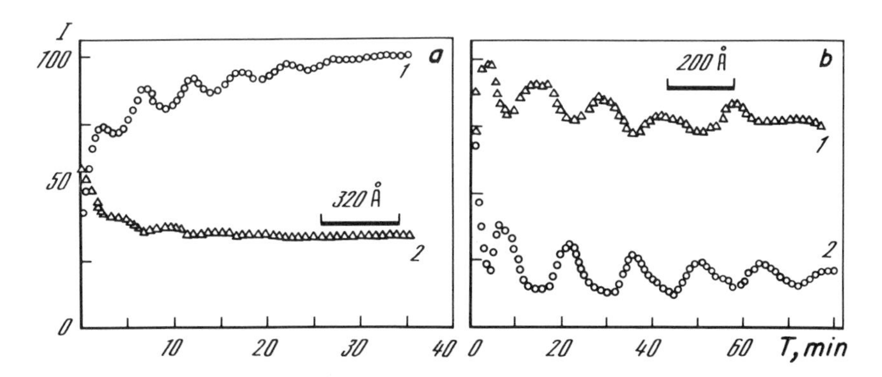

Figure 3.7. Relative intensities (1) of Ge (1) and Nb (2) Auger-lines from a specimen consisting of successive 100 Å Ge and Nb layers.
(a) Ar^+ ions with an energy of 1 KeV; (b) N_2^+ ions with an energy of 1 KeV. T = sputtering time, min.

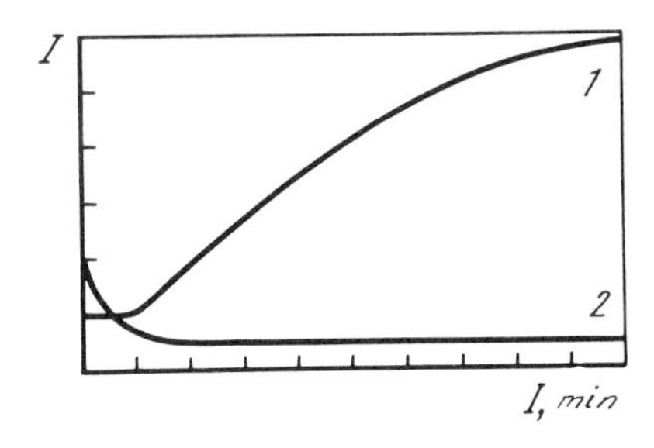

Figure 3.8. Theoretical dependence of the intensity of the nucleus (1) and surface (2) lines in spherical particles for sputtering time T.

Let us now examine the effect of diffusion stimulated by bombardment on changes in the initial distribution $n(z)$. Interesting results were obtained in study [35], which shows that Cu atoms in a surface layer of Si with a thickness of some 60 nm diffuse towards the surface under bombardment with Ar^+ ions. This diffusion is explained [35] by defects produced in the surface layer by bombardment. Reference [36] describes a very remarkable fact of the diffusion of Na^+ during ion bombardment of an SiO_2 implanted with Na^+ ions. The Na^+ diffusion was found to depend on the charge induced on the surface by bombardment. Under bombardment with O^- ions, Na^+ ions diffused towards the SiO_2 surface, whereas under bombardment with O^+ ions they accumulated at the SiO_2/Si interface, drifting away from the positively charged SiO_2 surface. These effects have a substantial role to play in studies of glass (see [37] and Section 5.6), and also of superionic conductors [38].

In view of Equations (3.6) and (3.10), it is worthwhile reviewing the results of study [39], which make it possible to account both for the statistical character of ion sputtering and for preferential sputtering within the model of sequential layer sputtering (SLS)—but not for the 'knock-in' effect.

In accordance with the SLS model, the contribution $\theta_n(t)$ of the nth layer to the surface layer sputtering for time t equals

$$\theta_n(t) = (t/t_0)/(n-1)! \cdot \exp(t/t_0) \tag{3.11}$$

where t is the time needed to remove one equivalent monolayer.

The intensity of the signal from the nth layer and the underlying k layers, with due account for the electron free path λ, equals

$$I_n(t) = I_0 \sum_{k=0}^{\infty} x_{n+k}\theta_n(t)\exp(-k/x) \tag{3.12}$$

where x_{n+k} is the concentration of the element being analysed in the kth layer. Measuring relative depth z in the $z = t/t_0$ monolayers on the basis of (3.11) and (3.12), we obtain

$$I_t = I_0 \sum_{k=0}^{M} \sum_{n=1}^{N} x_{n+k} \frac{z^{(n-1)}}{(n-1)!} \cdot \exp(-z)\exp(-k/x) \tag{3.13}$$

where M and N are the limits of practical summation by k and n with $I_{(Z)} \to 0$.

To account for the effect of ion bombardment, the value $x_n(t)$, obtained from preferential sputtering, must be substituted into the formula. The values $x_n(t) = f(z, t)$ are discussed in Section 3.3 [see e.g. (3.19)]. Generally speaking, it is also possible to account for the contamination surface layer [38]. Equations (3.11) and (3.12) are analogous to (3.6) and (3.10) for the continuous sputtering model. Equations (3.24)–(3.27) below supply data for I values with due regard for preferential sputtering upon the achievement of steady state in the concentration distribution.

Equations (3.10) and (3.13), however, do not make it possible to determine the true distribution of concentrations $C(x)$ by depth through the measured intensity of X-ray photoelectron lines $I(x)$. This problem—restoration of the concentration profile—was tackled in [21, 39–45]. In the general case, the values $C(x)$ and $I(x)$ are bound by an integral equation:

$$I_{(x)} = \int_{-\infty}^{+\infty} C(x) g(x - y, y) \mathrm{d}y \tag{3.14}$$

where $g(x - y, y)$ is the apparatus function accounting for the distortion of the true profile in ion sputtering. Equation (3.14) has to date been solved only for some of the simpler cases. Reference [41] describes a method of correcting experimental data on the basis of the assumption that the specimen surface remains absolutely smooth during sputtering while all distortions are explained by the limited depth of analysis L (depth of analysis L equals the electron mean free path λ multiplied by the sine of the exit angle $\lambda \sin \alpha$ for XPS and Auger spectroscopy). The term $g(x - y, y)$ merely equals $\exp(-y/L)$, and we obtain instead of (3.14):

$$I(x) = \int_0^{\infty} C(y) \exp(-y/L) \mathrm{d}y \tag{3.15}$$

The latter equation is easy to solve:

$$C(x) = I(x) - L \cdot [\mathrm{d}I(x)/\mathrm{d}x]. \tag{3.16}$$

Equation (3.16) was used in [41] to restore the profile at the SiO_2–Si, and in [42, 43] at the Ge–Si and NbO_2–Nb and at the interface between nickel, a thin film of chrome (1 nm) and nickel. The conclusion has been drawn that Equation (3.16) makes it possible to obtain a more correct location of the boundary dividing the phases. Moreover, the correction of the measured dependences $I(x)$ by the limited depth of analysis results in, first, the apparent profiles becoming steeper and, second, the crossing point of profiles for concentrations of different elements shifting deeper into the sample by $\approx 0.7.L$.

It should be noted that distortion of profiles due to the limited depth of analysis is not the main distortion factor. The most marked distortion effect is exerted on true profiles by the substantial roughness of the specimen surface, by the statistical character of ion sputtering, and also by the 'knock-in' effect and preferential sputtering. Regrettably, the known methods of solving the equation are only valid when the roughness of the specimen surface is taken into account

[21, 44, 45]. Depending on the adopted ion sputtering model, the apparatus function $g(x-y, y)$ may take different forms. But in view of the complex nature of the processes occurring in specimen sputtering by an ion beam, $g(x-y, y)$ is determined experimentally on the basis of changes in the concentration profile for the given layer structures. For instance, function $g(x, y)$ for a specimen thin layer of some element at depth y merely equals the measured dependence $I(x)$. The apparatus function can also be determined through measuring $I(x)$ for a film with thickness y deposited on a semi-infinite sublayer. In the latter case, $g(x, y)$ equals derivative $dI(x)/dx$. For some systems, (Ta_2O_5–Ta, SiO_2–Si), $g(x-y, y)$ turns out to be independent of y and therefore of sputtering time. In this case Equation (3.14) has a simpler form:

$$I(x) = \int_{-\infty}^{+\infty} C(x)g(x-y)\mathrm{d}y \tag{3.17}$$

and is effectively solved by the Fourier transform method [21, 44]. In the more general case, when the dependence of the apparatus function on the thickness of the sputtered layer cannot be disregarded, Equation (3.17) should be solved either through the interaction procedure suggested in [44] or through a method based on the series expansion of $I(x)$- and $C(x)$-functions, suggested in [45].

3.3. Ion sputtering and alloy analysis

As shown in Section 3.1, sputtering yields can differ essentially for various alloy components. For this reason we should expect—and actually find—that in ion sputtering the surface layer is enriched by the less easily sputtered components. Ion bombardment of Fe/Cr alloys, for example, enriches the surface layer with Fe [46]. In CuNi alloys the surface is enriched with Ni [47], in NiPd and AuPd alloys with Pd [48, 49], in AgAu with Au [50, 51], in AgCu with Cu [52], AgPd with Pd [53], and in Fe–Cr–Mo with Mo [54] (see also [55–61]). Enrichment of the alloy surface also depends on sputtering conditions. For instance, preferential sputtering of Fe in the FeNi alloy was discovered by some researchers, while others have not observed it (see [27]). These findings seem to be explained by the fact that enrichment of the surface by Ni atoms in the FeNi alloy is observed for a vertical ion beam, while for a grazing beam incidence, Fe enrichment occurs [62].

Surface enrichment by a less easily sputtered component occurs only at the beginning of ion sputtering, after which equilibrium is attained. Ratios of the atomic bulk concentrations C_1^v and C_2^v of the binary AB alloy and the surface concentrations C_1^s and C_2^s in equilibrium are considered in many papers (see, for instance, [47, 63, 64]).

Suppose that the sputtering yields of atoms A and B in the alloy equal γ_1 and γ_2, and that atoms are removed only from the surface layer. In this case the change in the concentration $C_1^s(t)$ of atom A with time can be expressed as

$$\mathrm{d}C_1^s/\mathrm{d}t = (-C_1^s\gamma_1 + C_1^s\gamma_1 C_1^v + C_2^s\gamma_2 C_1^v)\mathcal{T}, \tag{3.18}$$

where $\mathcal{T}$ is the ion flux per second per unit of area. The first term on the right of (3.18) represents the departure of A atoms (the characteristics of this atom are designated by the subscript 1) as a result of sputtering, the second describes the appearance of A atoms on the surface with probability C_1^v during the sputtering of atom A in the first monolayer. The third term describes the appearance of A atom on the surface with probability C_1^v during the sputtering of the atom B in the first monolayer. The steady state is characterized by the condition

$$dC_1^s/dt = 0.$$

Taking into account that $C_1^s + C_2^s = 1$, $C_1^v + C_2^v = 1$, we obtain:

$$\bar{C}_1^s/\bar{C}_2^s = (C_1^v/C_2^v)(\gamma_2/\gamma_1), \tag{3.19}$$

$$\bar{C}_1^s = \gamma_2 C_1^v/\gamma_1[I - C_1^v + (\gamma_2/\gamma_1)C_1^v] \tag{3.20}$$

where $\bar{C}_1$ is the concentration of the i th component on the surface under steady-state conditions.

Ratios analogous to (3.19) and (3.20) can be easily found for multi-component alloys. Thus, instead of (3.19), the following ratio holds (for any i and k):

$$\bar{C}_i^s/\bar{C}_k^s = (C_i^v/C_k^v)(\gamma_k/\gamma_i). \tag{3.21}$$

It is also possible to express C_i^{-s} as a function of C_i^v, C_k^v, γ_i, and γ_k.

The values C_i^v can be determined, i.e. an in-depth analysis of alloy components can be made, on the basis of formula (3.21), which also holds for the case when C depends on x (x being the distance from the surface). In this case it is necessary first to plot graduation curves of I_i/I_k vs C_i^v/C_k^v, where I is the intensity of the X-ray photoelectron line (Fig. 3.9 [65]). From (2.32), it follows that

$$I_i \approx \sigma_i \bar{C}_i^s \lambda_i.$$

Hence*,

$$I_i/I_k \sim (\sigma_i\lambda_i/\sigma_k\lambda_k)(C_i^v/C_k^v)(\gamma_k/\gamma_i). \tag{3.22}$$

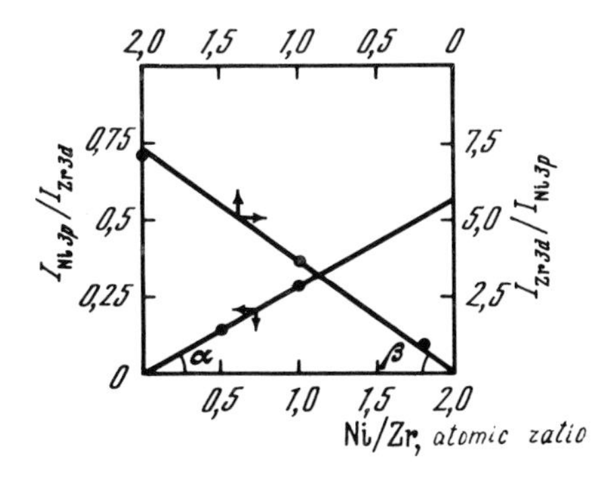

Figure 3.9. Graduation curve for Zr/Ni alloys.

*Formula (3.22) is valid only in the case of constant concentration C_1^s in the altered layer within thickness δ, if $\delta > \lambda$. In the general case the ratio I_i/I_k may depend on C_i^s/C_k^s in a more complicated manner [see, for instance, (3.31)].

If γ_k/γ_i is independent of the alloy composition (as is often the case), the values I_i/I_k linearly depend on C_i^v/C_k^v (Fig. 3.9). In this case the value γ_i/γ_k can be determined by using the values σ_i, λ_i, σ_k, λ_k and the slope of the graduation curve.

Note that the values γ_i/γ_k in alloys do not necessarily coincide with this ratio in pure metals. A difference between absolute values of γ for metals and for alloys is also possible [66]. For example, γ_{Cu} is larger in Cu_3Au than in the pure metal [67], the alloy sputtering yield being greater than that of any of its components. In the AgAu alloy the ratio $\gamma_{Ag}/\gamma_{Au} = 2$, whereas for pure metals it is close to unity [68]. The values γ for Cr and for Au in the alloy and in pure metals are compared in paper [69] it some detail. The ratio γ_{Cr}/γ_{Au} in the alloy at $C_{Cr} = 1\%$ equals 0.5 and increases to 1 at $C_{Cr} = 20\%$. The ratio γ_{Cr}/γ_{Au} depend only slightly on the energy (0.5–2 keV) and type (Ne^+, Ar^+) of bombarding ions. The value γ_{Au} is 7.9 for pure gold (Ar^+, $E_{kin} = 2$ keV) but in the alloy AuCr, γ_{Au} is 5 at with 10–20% Cr. The value $\gamma_{Cr} = 5$ in this alloy depends slightly on composition and is about 2.5 times that pure chrome. It is shown in paper [55] that Al and Si in AlPd and SiPd alloys are sputtered more easily than Pd, although γ_{Pd} is greater than γ_{Al} and γ_{Si} in solid elements.

If γ_i/γ_k depends on concentration, graduation curves enable us to obtain this ratio as a function of the alloy composition. Regardless of their type, graduation curves make it possible to transform the dependence of intensities I on sputtering time to the dependence of C_i^v on sputtering time (or distance from the surface).

The development of Equations (3.20), (3.21) was based on the assumption that ion sputtering affects only the surface monolayer. In reality a surface layer of thickness δ is altered. Assuming that concentration C_1^s in this altered layer is constant, the following equation, similar to (3.18) [70], can be written;

$$dN\delta C_1^s/dt = \mathcal{T}[\gamma_1 C_1^s C_1^v + \gamma_2(1 - C_1^s)C_1^v - \gamma_1 C_1^s], \tag{3.23}$$

where N is the mean atomic density of the alloy. Equations (3.20) and (3.22) can be obtained for the steady state.

For the dependence of the value C_1^s one time, we have:

$$dN\delta C_s^1/dt \approx N\delta dC_1^s/dt, \quad dC_1^s/dt + C_1^s/\tau = BC_1^v \tag{3.24}$$

where

$$1/\tau = \mathcal{T}/\delta N[\gamma_1 - (\gamma_1 - \gamma_2)C_1^x]_i \quad B = \mathcal{T}\gamma_2/\delta N.$$

Solving (3.24) under boundary conditions $C_1^s = C_1^v$ and $t = 0$ yields

$$C_1^s(t) = (C_1^v - B\tau C_1^v)\exp(-t/\tau) + B\tau C_1^v, \tag{3.25}$$

or

$$C_1^s(t) = (C_1^v - \bar{C}_1^s)\exp(-t/\tau) + \bar{C}_1^s.$$

Equation (3.25) is also valid for the case when C_1^v is a function of x, where x is the distance from the surface.

To determine the value δ within the framework of this model, the following formula can be used

$$\exp(-t/\tau) = (R_t - R_\infty)/\{R_0 - R_\infty[R_0(1 - C_1^v) + R_t C_1^v]\} \tag{3.26}$$

where R_i is the ratio C_1^s/C_2^s at time i. Formula (3.26) readily yields τ, while δ can be found from (3.24).

Determination of δ [70] by this method for the CuNi alloys showed that δ depends slightly on composition and is equal roughly to 1.0 nm at $E_{kin} = 0.5$ keV for Ar^+ ions. When E_{kin} for Ar^+ rises from 0.5 to 2 keV the value δ in the CuNi alloy (Cu = 38.4%) grows from 1.0 to 2.5 nm. The value δ can be estimated also using the decrease in the value $C_1^s/(C_2^s + C_1^s)$ for superthin films, where subscript 1 refers to the easily sputtered component [71]. (For some other methods of determining δ see reference [71].) Estimates of δ for other alloys also yield values whose order of magnitude is several dozen nm [71]: 1.0–5.0 nm in CuNi, 3.0–4.0 nm in Cu_3Au, 4.0–5.0 nm in Ge, 1.5–3.0 nm in AgAu. The value δ is strongly model-dependent. The value 53.0 nm was obtained for δ in Pt_2Si in [55], this noticeably exceeding the other values cited above. According to the theoretical analysis reported in [72], the constant concentration C_1^s corresponding to the altered layer in the steady state is already attained after the removal, on average, of one monolayer.

Reference [70] considers the extent to which δ corresponds to the value y characterizing the layer which has to be removed in order to attain the steady state:

$$y = \tau V,$$

where τ is obtained from (3.24) while V is the sputtering speed: $V = \mathscr{T}/N[\gamma_1 C_1^s + \gamma_2(1 - C_1^s)$ For the relationship $L = \delta/\tau V$ we obtain

$$L = [\gamma_1 - (\gamma_1 - \gamma_2)C_1^v]/[\gamma_2 - (\gamma_2 - \gamma_1)C_1^s] \tag{3.27}$$

Normally $\gamma_1 \approx \gamma_2$, so $L \sim 1$. But if $C_1^v \ll 1$, $\gamma_1 \gg \gamma_2$, then $L = \gamma_1/\gamma_2 \gg 1$, i.e. the steady state is achieved when removing a layer having a thickness less than the value δ. If $C_1^v \ll 1$, $\gamma_2 \gg \gamma_1$, then $L < 1$.

The above models do not take account of the influence of diffusion on the composition of the surface layer. Diffusion is taken into account in models worked out in papers [73–77]. Note that allowance for diffusion does not influence relationships (3.20) and (3.21), characteristic of the steady state*, but the notion of the nature of the altered layer changes. According to Ho's model [73], the composition of the altered layer in the steady state changes as follows

$$C_1^s(x) = C_1^v[1 - H\exp(-xV/D)/(H - 1)]. \tag{3.28}$$

Here V is the sputtering speed; D is the diffusion coefficient of the A atom under the conditions of ion bombardment, and x the distance from the surface:

$$H = \{(\gamma_2 - \gamma_1)[1 - C_1(0,t)]\}/\{\gamma_1 C_1(0,t) + \gamma_2[1 - C(0,t)]\}$$
$$C_1(0,0) = C_1^v;\ C_1 = (0,\infty) = \bar{C}_1^s.$$

Differentiating (3.28) with respect to x, we obtain

$$(\bar{C}_1^s - C_1^v)/\delta = (\delta\bar{C}_1^s/\mathrm{d}x)_{x=0} \qquad \text{where } \delta = D/V. \tag{3.29}$$

The value δ can by definition be accepted as the width of the altered layer. The

*Taking the diffusion into consideration influences the nature of attainment of the steady state, i.e. on the change of C_1^s with time.

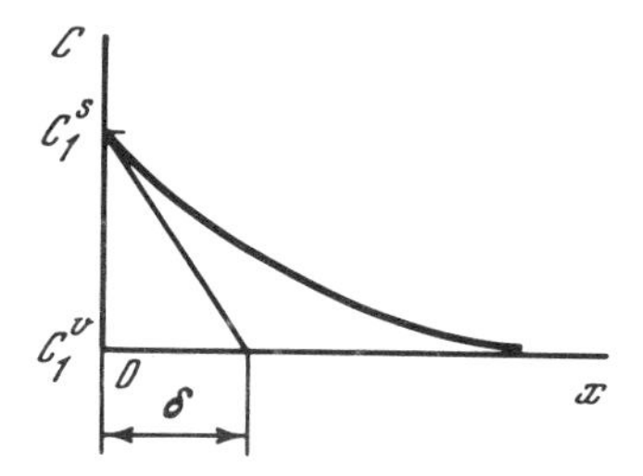

Figure 3.10. Determination of thickness δ of the transition layer.

physical meaning of this difinition of δ is graphically represented in Fig. (3.10). Taking into account Equations (3.22), (2.55), and (2.56), the intensity of the line I_i from the altered layer is

$$I_1 \approx \delta_1 \lambda_1 C_1^v [1 - H\delta/(H-1)(\delta + \lambda \sin \alpha)]. \tag{3.30}$$

Substituting the value of H into (3.30), we obtain

$$I_1/I_2 = (\sigma_1 \lambda_1 C_1^v / \sigma_2 \lambda_2 C_2^v)(\gamma_2/\gamma_1) \times \frac{(\delta + \lambda_2 \sin \alpha)(\gamma_1 \lambda_1 \sin \alpha + \delta \gamma_1 \bar{C}_1^s + \delta \gamma_2 C1_2^s)}{(\delta + \lambda_1 \sin \alpha)(\gamma_2 \lambda_2 \sin \alpha + \delta \gamma_1 \bar{C}_1^s + \delta \gamma_2 \bar{C}_2^s)}, \; C_i^s(0) \equiv \bar{C}_i^s. \tag{3.31}$$

If $\delta \gg \lambda$, then (3.31) is equivalent to (3.22). If $\lambda \gg \delta$ or $\delta \sim 0$, then

$$I_1/I_2 = \sigma_1 \lambda_1 C_1^v / \sigma_2 \lambda_2 C_2^v. \tag{3.32}$$

In the general case of $\lambda \sim \delta$ we can write

$$I_1/I_2 = \sigma_1 \lambda_1 C_1^v / \sigma_2 \lambda_2 C_2^v (\gamma_2/\gamma_1)^k \tag{3.33}$$

where $0 \leqslant k \leqslant 1$.

Results on the concentration variations in the altered layer that were obtained in [75] are somewhat different from those obtained in [73] (see (3.28)). Figure (3.11a, b) represents variations in the concentration of a component which is sputtered more easily in the altered layer, as they were determined in [73, 75]. In paper [51] distribution of copper concentrations in the altered layer of the CuNi alloy and of Ag concentrations in the AgAu alloy are determined by methods of Auger electron spectroscopy (AES) and secondary ion mass spectroscopy (SIMS). The profile represented in Fig. (3.11c) was obtained in [51]. The value γ_{Cu}/γ_{Ni} in paper [51] is considered to be roughly equivalent to unity [derived from formula (3.19), which always holds for the topmost monolayer]. According to findings of study [51], it is only in the region II (see Fig. 3.11) that the change of concentration depends on the diffusion process, whereas in region I the topmost layer is enriched with copper atoms as a consequence of copper segregation, because copper has a lower surface bonding energy than nickel. The authors of paper [51] believe that the composition of the topmost part of the altered layer depends on thermodynamic segregation processes. Their view is supported by the dependence of the concentration profile in the altered layer on temperature. Since it is elements with low surface bonding energies [78] that are prone to segregation

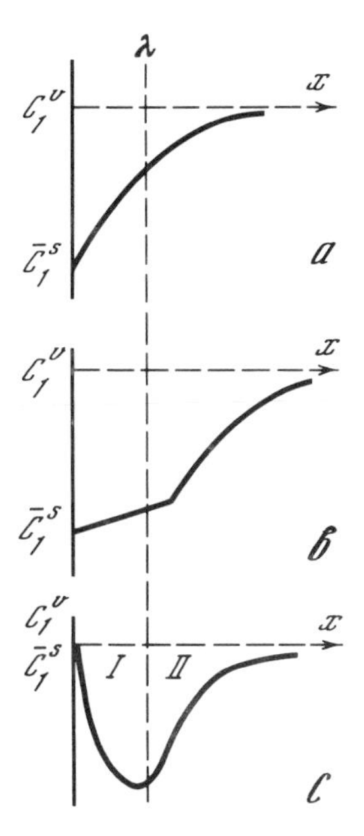

Figure 3.11. Different versions of change in concentration C_1^s of the easily sputtered component of the alloy in the modified layer depending on depth x.

and preferential sputtering, the curve shown in Fig. (3.11c) should hold for all the easily sputtered elements. The profile of the concentration of Au in the modified layer of $Au_{0.56}Cu_{0.44}$ alloy is close to that shown in Fig. (3.11b). The minimum disappears with heating [79]. Incidentally, measuring the slope of the X-ray photoelectron line intensities for individual components in the altered layer would make it possible unambiguously to distinguish profiles represented in Fig. (3.11a, b) from those in Fig. (3.11c).

The relationship between C_i^s and C_i^v therefore depends on γ_i, γ_k and at that the values of γ_i and their ratios may be different in metal and alloy and also may depend on alloy concentration. The dependence of γ_i^a on C_i^v and γ_i^m (a stands for alloy and m for metal) was considered in [80]. The following formula was suggested in that paper:

$$\gamma_1^a(C_1^v) = [C_1^v/u_1^a(C_1^v)](u_1^m\gamma_1^m C_1^v + C_2^v u_2^m \gamma_2^m), \tag{3.34}$$

where u is the surface bonding energy of an atom. This formula qualitatively explains some experimental factors but is difficult to verify because of the lack of values for u. The same author suggests in a later paper [57] a different approach to the ratios γ_1^a/γ_2^a in alloys. This model presumes that the energy of primary ions is quickly distributed between all the atoms and ions in a cascade of collisions caused by primary ions. This assumption may be wrong for atoms that differ considerably in mass but it explains why the surface becomes enriched with heavy atoms fairly often regardless of the energy and mass of the primary ion.

Let us consider the mean free path L of the atome A in a cascade during the time

τ of its existence. Assuming that this movement is random, we obtain

$$L \sim (vl\tau)^{1/2}, \tag{3.35}$$

where v is mean velocity, and l the mean free path of an atom or ion in the cascade. If furthermore, the relative sputtering probability p_i is $L/(L_1 + L_2) = (1 + L_2/L_1)^{-1}$ and since $E_{\text{kin}} = \text{mv}^2/2$ and $L \sim v^{1/2}$ (see (3.35)), then

$$L_1/L_2 = (m_1/m_2)^{1/4}. \tag{3.36}$$

Since $p_i = \gamma_i/(\gamma_1 + \gamma_2)$, then using (3.20) we obtain

$$\bar{C}_s^1 = x_1/(1 - x_1), \tag{3.37}$$
$$x_1 = C_1^v/C_2^v(m_1/m_2)^{1/4}.$$

Formula (3.37) effectively explains a number of experimental data (Fig. 3.12 [57]).

Paper [81] suggests a different formula to take account of the dependence of γ^a on m and on the mass m_i of the ion used for sputtering:

$$\gamma_1^a/\gamma_2^a = (m_1/m_2)(m_i + m_1)^2/(m_i + m_2)^2. \tag{3.38}$$

This formula correctly represents changes of the ratio $\gamma_{\text{Ga}}/\gamma_{\text{As}}$ in GaAs and $\gamma_{\text{In}}/\gamma_{\text{Sb}}$ in InSb in the case of bombardment with Ar^+, Kr^+, and Xe^+ ions. These changes, however, are very insignificant. Equations (3.35), (3.37), and (3.38) seem to be applicable within very narrow bounds: they cannot explain, for instance, the preferential sputtering of the heavier Ag atom in AgCu [52] and the complex change of γ_{Cr} and γ_{Cu} in CrCu [69].

A detailed analysis of the interdependence of ratios for components of binary alloys and compounds and such ratios for free elements are supplied in [82]. Table 3.4 [82] gives a list of experimental values of γ_A/γ_B in alloys bombarded by Ar^+ ions (usually with $E_{\text{kin}} = 2$ keV). A comparison with the value of γ_A/γ_B for free elements, computed for bombardment with Ar^+ ($E_{\text{kin}} = 2$keV) according to theory [83], shows that there is qualitative agreement, though of a low order, between ratio γ_A/γ_B in alloys and in free metal in most cases (with the exception of the six bottom lines in Table 3.4). This is explained [82] by the fact that in these cases the ratio of atomic weights m_B/m_A is close to unity and the role is played by the surface bonding energy of the atom, which is roughly equal for atoms in

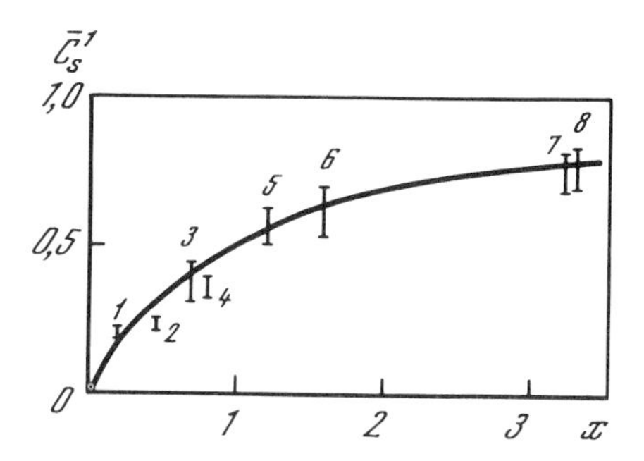

Figure 3.12.

Table 3.4
Relationship of γ_A/γ_B values in an alloy with atomic mass m_B/m_A and γ_A/γ_B for free elements

A – B[a] system	m_B/m_A	γ_A/γ_B (alloy)	γ_A/γ_B (element)
Ag–Au	1.8	1.7–1.8	1.32
$Ag–Au_{0.19}$	1.8	1.2	1.32
Cu–Au	3.1	1.0	1.04
Cu_3Au	3.1	1.1	1.04
Ag–Pd	1.0	2.2–2.7	1.38
CuPd	0.5	1.0–1.4	1.04
CuPt	3.1	1.6–1.3	1.5
NiPt	3.3	1.5–1.9	1.17
CuNi	0.9	1.7	1.29
PdNi	0.6	1.5–1.7	1.29
UNb	0.4	1.7	1.43
Au–Cr	0.3	1.4–2.0	1.34
$Al–Au_2$	7.3	1.9	0.57
Al_2Au	7.3	1.3	0.53
Si–Pt	7.0	2.1	0.67
$Si–Pt_2$	7.0	1.6	0.67
Si–Ni	2.1	1.6	0.57
Al–Cu	2.4		0.60

[a]Element B accumulates on the surface during ion sputtering.

metal and in alloys. That is why the ratios γ_A/γ_B in alloys and in metal are close. Conversely, the difference between the masses m_A and m_B for the six bottom systems is considerable, and it is this difference that determines the enrichment of the surface with the heavy component of the system in the process of bombardment. References [84, 85] review changes induced in the composition of multicomponent systems by ion bombardment.

3.4. Changes in chemical compounds under the effect of ion sputtering

Changes in the composition of the surface layer occur as a result of ion sputtering. Although chemical bonding energy is several orders below the energy of ions ($\sim$ 1–10 keV) usually used for sputtering, changes in the chemical composition cannot be described as the preferential sputtering of certain atoms but should be viewed as a result of chemical reactions induced by ion bombardment [86, 87]. For instance, $CuCl_2$ is easily reduced by sputtering to CuCl, while further reduction to Cu is rather difficult. This also is indicated by the dependence of ease with which oxides are reduced on the standard heat of formation ΔG (Table 3.5) [88]: the lower the ΔG value, the easier it is to reduced oxides to metal.

Other examples of reduction reactions under the effect of ion bombardment are the reduction of oxides and halides to metal [86, 88–94], K_2PtCl_6 to K_2PtCl_4 [95], the enrichment of the surface layer of GaAs and GaP with metallic Ga [96], the disintegration of FeS and FeS_2 into sulphur and metallic iron [87], and reduction of $CuSiF_6$ to Cu [97]. The available data suggest the following conclusions:

Table 3.5
Oxide sputtering characteristics

Oxide	G^a	Reduction	Oxide	G^a	Reduction
Au_2O_3	−39	Yes	$Ni(OH)_2$	108	No
AgO	2.6	Yes	MoO_2	119	No
PdO	20	Yes	SnO_2	124	No
CuO	30	Yes	MoO_3	162	Yes
Cu_2O	35	Yes	Fe_2O_3	177	Yes
PbO	45	Yes	SiO_2	192	No
NiO	52	Yes	Cr_2O_3	249	No
PbO_2	52	Yes	Ti_2O_3	346	No
FeO	58	Yes	Al_2O_3	377	No
RuO_2	60	Yes	Ta_2O_5	471	No

[a]Standard heat of formation in kcal.

1. The extent of reaction stands in proportion to the dosage of ion bombardment [87]. Higher densities of ionic current also facilitate the process in a number of cases [86].
2. A rise in the ion E_{kin} accelerates reactions.
3. Reduction proceeds faster when the ion beam falls at right angles to the surface.
4. The speed of reduction depends in a number of cases on the nature of the ions.
5. Thin oxide layers on metal are more easily reduced than compact oxides. In particular, thin SiO_2 films on Si and Ta_2O_5 films on Ta can be reduced [86, 98] while the reduction of compact oxides is difficult. This seems to be explained by the possible 'knock-in' of oxygen into the sublayer and its subsequent fixation.

It follows from conclusions 2 and 3 that reduction takes place in the layer penetrated by ions since higher E_{kin} and the right angle of fall increase the ion path in a solid.

There also are, along with reduction, oxidation phenomena under the effect of ion bombardment, for example $NO_2^- \rightarrow NO_3^-$ [93], and the formation of oxide on the surface of Mn alloys with unusually low partial oxygen pressures [99]. Organic compounds disintegrate especially fast under ion bombardment [93]. Chemical reactions between different compounds in the sample can also be induced by ion bombardment. For instance, [86], C→F are formed bonds in the bombardment of a mixture of carbon and Na_2SiF_6, while manganese and zirconium carbides form where Mn and Zr surfaces contaminated with carbon compounds are bombarded. Carbides, sulphides, and nitrates of elements of Groups II and IV become oxides [100]. Also, when the ions of molecules or atoms capable of forming chemical bonds with atoms of the sample are used for ion bombardment, surface compounds can be formed which drastically alter the composition of the surface. (See e.g., [100], which examines the interaction of N_2^+ and NO^+ with graphite, and [101], which describes the formation of nitrides in the interaction of N_2^+ with some oxides of elements of Group IV.)

All this gives us reason to conclude that ion bombardment cannot be used for cleaning or layer-by-layer sputtering of many inorganic and most organic compounds. It is necessary in every case to study specifically the stability of compounds under the effect of ion bombardment and the compatibility of the possible artefacts of ion sputtering with the tasks of a given experiment.

References

1. Elfstrom, B. O. and Olefford, I. (1977) Preparation of alloys for ESCA investigation. *Phys. Scr.* **16**, 436–441.
2. McCracken, G. M. (1975) Behaviour of surface under ion bombardment. *Rep. Progr. Phys.* **38**, 241–280.
3. Dearnaley, G., Freeman, J. H., Nelson, R. S., and Stephen, J. (1973) *Ion implantation. North-Holland, Amsterdam.*
4. Arifov, U. A. (1968) *Interaction of atomic particles with the surface of solids.* Nauka, Moscow (in Russian).
5. Wenner, G. (1979) Ion sputtering and surface analysis. In: *Methods of surface analysis*, Mir, Moscow, pp. 18–59 (in Russian).
6. Oechsner, H. (1975) Sputtering: review of some recent experimental and theoretical aspects. *Appl. Phys.* **8**, 185–198.
7. Coburn, J. W. (1976) Sputtering in surface analysis of solids—discussion of some problems. *J. Vac. Sci. Technol.* **13**, 1037–1044.
8. Behrisch, R., Ed. (1983) *Sputtering by particle bombardment. Topics in Applied Physics*, Vol. 52, Springer, Berlin.
9. Hawkins, D. H. (1979) Ion milling (ion beam etching), 1975–1978: A bibliography. *J. Vac. Sci. Technol.* **16**, 1051–1071.
10. Musket, R. G., McLean, W., Colmenares, C. A. *et al.* (1982) Preparation of atomic clean surfaces of selected elements. A review. *Appl. Surf. Sci.* **10**, 143–207.
11. Nefedov, V. I. (1981) X-ray photoelectron analysis of surface layers with concentration gradient. *Surf. Interface Anal.* **3**, 72–75.
12. Kelly, R. and Lam, N. Q. (1973) The sputtering of oxides. A survey of the experimental results. *Radiat. Eff.* **19**, 39–48.
13. Oechsner, H. (1973) Sputtering of polycrystalline metal surfaces of oblique ion bombardment in 1 keV range. *Z. Phys.* **261**, 37–58.
14. Hosaka, S. and Hashimoto, S. (1978) Influence of sample inclination and rotation during ion-beam etching on ion-etched structures. *J. Vac. Sci. Technol.* **15**, 1712–1717.
15. Kelly, R. (1979) Thermal effects in sputtering. *Surf. Sci.* **90**, 280–318.
16. Prival, H. G. (1978) Model of ion sputtering process. *Surf. Sci.* **76**, 443–463.
17. Hoffman, D. W. (1975) A cratering analysis for quantitative depth profiling by ion beam sputtering. *Surf. Sci.* **50**, 29–52.
18. Bradley, L., Bosworth, J. M., Briggs, D. *et al.* (1978) Uniform depth profiling in X-ray photoelectron spectroscopy. *Appl. Spectrosc.* **32**, 175–177.
19. Hofmann, S. (1976) Evaluation of concentration depth profiles by sputtering in SIMS and AES. *Appl. Phys.* **9**, 59–66.
20. Coburn, J. W. and Winters, H. F. (1979) Plasma etching—a discussion of mechanism. *J. Vac. Sci. Technol.* **16**, 392–403.
21. Hofmann, S. (1980) Quantitative depth profiling in surface analysis. A review. *Surf. Interface Anal.* **2**, 148–166.

22. Castle, J. E. and Nuzel, L. B. (1977) A nomographic method for identification of interface position on ion etch profiles. *J. Electron Spectrosc. Rel. Phen.* **12**, 195–202.
23. Benninghoven, A. (1970) Analysis of monomolecular layers of solids by secondary ion emission. *Z. Phys.* **230**, 403–420.
24. Hofer, W. O. and Martin, P. J. (1978) Influence of reactive gases on sputtering and secondary ion emission oxidation of titanium and vanadium during energetic particle irradiation. *Appl. Phys.* **16**, 271–278.
25. Mathieu, H. J., McClure, D. E., and Landolt, D. (1976) Influence of ion bombardment on depth resolution in Auger electron spectroscopy analysis of thin gold films on nickel. *Thin Solid Films* **38**, 281–294.
26. Shimizu, R. (1979) Comment on 'Evaluation of concentration depth profiles by sputtering in SIMS and AES' by Hofmann, S. *Appl. Phys.* **18**, 425–426.
27. Nefedov, V. I., Pozdeev, P. P., Dorfman, V. E. and Pypkin, B. N. (1980) X-ray photoelectron study of thin evaporated films of Fe/Ni alloy. *Surf. Interface Anal.* **2**, 26–30.
28. Liau, Z. L., Tsaur, B. Y., and Mayer, J. W. (1979) Influence of atomic mixing and preferential sputtering on depth profiles and interfaces. *J. Vac. Sci. Technol.* **16**, 121–127.
29. Schulz, F., Wittmaack, K., and Maul, J. (1973) Implication in the use of secondary ion mass spectrometry to investigation impurity concentration profiles in solids. *Radiat. Eff.* **18**, 211–215.
30. Kang, S. T., Shimizu, R., and Okutani, T. (1979) Sputtering of Si with keV Art ions. II. Computer simulation of sputter broadening due to ion bombardment in depth profiling. *Jap. J. Appl. Phys.* **18**, 1987–1994.
31. Ishitani, T. and Shimizu, R. (1975) Computer simulation of atomic mixing ion bombardment. *Appl. Phys.* **6**, 241–248.
32. Tarng, M. L. and Wehner, G. K. (1972) Auger electron spectroscopy studies of sputter deposition and sputter removal of Mo from various metal surfaces. *J. Appl. Phys.* **43**, 2268–2277.
33. Blattner, R. J., Nadel, S., Evans, C. A. *et al.* (1979) Improved depth resolution in auger depth profiling of multilayered thin films by reactive ion sputtering. *Surf. Interface Anal.* **1**, 32–35.
34. Cross, Y. M. and Dewing, J. (1979) Thickness measurements on layered in powdered form by means of XPS and ion sputtering. *Surf. Interface Anal.* **1**, 26–31.
35. Hart, R. R., Dunlap, H. L., and Marsh, O. J. (1975) Ion induced migration of Cu into Si. *J. Appl. Phys.* **46**, 194–1951.
36. Hughes, H. L., Baxter, R. D., and Phillips, B. F. (1972) Dependence of MOS device radiation sensitivity of oxide impurities. *IEEE Trans. Nucl. Sci.* **19**, 256–263.
37. Kleshchevnikov, A. M. (1983) In-depth profiles of elements concentrations in solids on the basis of X-ray photoelectron spectroscopy data. Synopsis of thesis for the degree of candidate of sciences. Moscow (in Russian).
38. Livshits, A. and Polak, M. (1982) Electron induced segregation to surfaces of the superionic conductor Na-alumina studied by AES and XPS. *Surf. Sci.* **119**, 314–330.
39. Hofmann, S. and Sanz, J. M. (1984) Quantification of preferential sputtering and contamination overlayer effects in AES sputter profiling. *Surf. Interface Anal.* **6**, 78–81.
40. Sanz, J. M. (1984) Cross correlation versus convolation. *Surf. Interface Anal.* **6**, 196.
41. Schwarz, S. A., Helms, C. R., Spicer, W. E., and Taylor, N. J. (1978) High resolution Auger sputter profiling study of the effect of phosphorus pile up on the Si–SiO_2 interface morphology. *J. Vac. Sci. Technol.* **15**, 227–230.

42. Etzkorn, H. W. and Kirschner, J. (1980) Depth resolution of sputter profiling investigated by combined auger-X-ray analysis of thin films. *Nucl. Instrum. Methods* **168**, 395–398.
43. Hofmann, S. (1979) Surface and thin film analysis: concepts, capabilities and limitations. *Talanta* **26**, 665–673.
44. Ho, P. S. and Lewis, J. E. (1976) Deconvolution method for composition profiling by Auger-sputtering technique. *Surf. Sci.* **55**, 335–448.
45. Palacio, C. and Martinez-Duart, J. M. (1983). Deconvolution methods applied to sputter depth profiles at interfaces. *Thin Solid Films* **105**, 25–32.
46. Holm, R. and Storp, S. (1976) Methods of obtaining in-depth data in surface analysis. *Vak. Techn.* **25**, 41–47.
47. Shimizu, H., Ono, H. and Nakayama, K. (1973) Quantitative Auger analysis of copper nickel alloy surfaces after Ar ion bombardment. *Surf. Sci.* **36**, 817–821.
48. Garbassi, F. and Parravano, G. (1978) Surface composition and oxygen chemisorption of Ag – Pd alloys. *Surf. Sci.* **71**, 42–50.
49. Slusser, G. J. and Winograd, N. (1979) Surface segregation of PdAg alloys induced by ion bombardment. *Surf. Sci.* **84**, 211–222.
50. Jabumoto, M., Watanabe, K. and Jamashita, T. (1978) AES study of surface segregation of Ag–Au alloys with ion bombardment and annealing. *Surf. Sci.* **77**, 615–625.
51. Jabumoto, M., Kakibayashi, H., Mohri, M. *et al.* (1979) Ag–Au alloys modified by ion bombardment. *Thin Solid Films* **63**, 263–268.
52. Braun, P. and Farber, W. (1975) AES studies of surface composition of Ag–Cu alloys. *Surf. Sci.* **47**, 57–63.
53. Jablonski, A., Overbury, S. H., and Somorjai, G. A. (1978) The surface composition of the gold–palladium binary alloy system. *Surf. Sci.* **65**, 578–592.
54. Mathieu, H. J. and Landolt, D. (1979) Influence of sputtering on the surface composition of Fe–Cr–Mo alloys. *Appl. Surf. Sci.* **3**, 348–355.
55. Ho, P. S., Lewis, J. E., and Chu, W. K. (1979) Preferred sputtering on binary alloy surfaces of the Al–Pd–Si system. *Surf. Sci.* **85**, 19–28.
56. Tompkins, H. G. (1979) Preferential sputtering in gold–nickel and gold–copper alloys. *J. Vac. Sci. Technol.* **16**, 778–780.
57. Haff, P. K. (1977) Model for surface layer composition changes in sputtered alloys and compunds. *Appl. Phys. Lett.* **31**, 259–266.
58. McGuire, G. E. (1978) Effects of ion sputtering on semiconductor surfaces. *Surf. Sci.* **76**, 130–147.
59. Frankenthal, R. P. and Siconolfi, D. J. (1982) Effect of ion sputtering on surface composition of binary alloys of tin, lead, and indium. *J. Vac. Sci. Technol.* **20**, 515–516.
60. Holloway, P. H. and Bhattacharya, R. S. (1982) Preferential sputtering of PtSi, $NiSi_2$, and AgAu. *J. Vac. Sci. Technol.* **20**, 444–448.
61. Taglauer, E. (1982) Surface modifications due to preferential sputtering. *Appl. Surf. Sci.* **13**, 80–93.
62. Olson, R. R. and Wehner, G. K. (1977) Composition variation as a function of ejection angle in sputtering of alloys. *J. Vac. Sci. Technol.* **14**, 319–321.
63. Patterson, W. L. and Shirn, G. A. (1967) Sputtering of nickel–chromium alloys. *J. Vac. Sci. Technol.* **4**, 343–347.
64. Holland, L. and Priestland, C. R. (1972) Influence of sputtering and transport mechanism on target etching and thin film growth in RF-system. *Vacuum* **22**, 133–136.

65. Nefedov, V. I., Chulkov, N. G. and Lunin (1980) Dependence of surface composition of Zr–Ni intermetallic compounds. *Surf. Interface Anal.* **2**, 207–211.
66. Sartwell, B. D. (1979) Thin film sputtering yields for Fe, Cr and FeCr alloy measured by PIXE. *J. Appl. Phys.* **50**, 7887–7893.
67. Ogar, W. T., Olson, N. T. and Smith, H. P. (1969) Simultaneous measurement of sputtering constituents of Cu_2Au. *J. Appl. Phys.* **40**, 4997–4999.
68. Ho, P. S., Lewis, J. E., and Howard, J. K. (1977) Auger study of preferred sputtering of Ag–Au alloy surfaces. *J. Vac. Sci. Technol.* **14**, 322–325.
69. Holloway, P. H. (1977) Quantitative Auger-electron analysis of homogeneous binary alloys: chromium in gold. *Surf. Sci.* **66**, 479–494.
70. Ho, P. S., Lewis, J. E., Wildman, H. S., and Howard, J. K. (1976) Auger study of preferred sputtering on binary alloy surfaces. *Surf. Sci.* **57**, 393–405.
71. Goto, K., Koshikawa, T., Ishikawa, K. and Shimizu, R. (1978) Preferential sputtering of coevaporated Cu–Ni films associated with altered layer. *Surf. Sci.* **75**, L373–L375.
72. Werner, H. W. and Warmholtz, N. (1976) Influence of selective sputtering on surface composition. *Surf. Sci.* **57**, 706–714.
73. Ho, P. S. (1978) Effects of enhanced diffusion on preferred sputtering of homogeneous alloy surfaces. *Surf. Sci.* **72**, 253–263.
74. Pickering, H. W. (1976) Ion sputtering of alloys. *J. Vac. Sci. Technol.* **13**, 618–621.
75. Arita, M. and Someno, M. (1977) Preferential sputtering on binary alloys by SIMS. Proceedings of the Seventh International Vacuum Congress and the Third International Conference on Solid Surfaces, Vienna, pp. 2511–2515.
76. Collins, R. (1979) Note on the time constant for preferential binary sputtering. *Radiat. Eff. Lett.* **43**, 111–116.
77. Webb, R., Carter, G., and Collins, R. (1978) Influence of preferential enhanced diffusion on composition changes in sputtering binary solids. *Radiat. Eff.* **39**, 129–139.
78. Brongersma, H. H., Spornaay, M. J. and Buck, T. M. (1978) Surface segregation in Cu–Ni, Cu–Pt alloys: comparison of low energy ion scattering results with theory. *Surf. Sci.* **71**, 657–678.
79. Li, R., Koshikawa, T., and Gotto, K. (1982) Changes in gold concentration at the surface of a Au–Cu alloy sputtered at low temperature. *Surf. Sci.* **121**, L561–L568.
80. Haff, P. K. and Switkowski, Z. E. (1976) Sputtering of binary compounds. *Appl. Phys. Lett.* **29**, 549–551.
81. McGuire, G. E. (1978) Effects of ion sputtering on semiconductor surfaces. *Surf. Sci.* **76**, 130–147.
82. Betz, G. (1980) Alloy sputtering. *Surf. Sci.* **92**, 283–309.
83. Sigmund, P. (1969) Theory of sputtering. I. Sputtering yield of amorphous and polycrystalline targets. *Phys. Rev.* **184**, 383–386.
84. Wehner, G. K. (1984) Sputtering of multicomponent materials. *J. Vac. Sci. Technol.* **A1**, 487–490.
85. Kowalski, Z. W. (1984) A review of compositional changes of multicomponent materials (especially biomaterials) induced by ion sputtering. *J. Vac. Sci. Technol.* **A1**, 494–496.
86. Holm, R. and Storp, S. (1977) ESCA studies on changes in surface composition under ion bombardment. *Appl. Phys.* **12**, 101–112.
87. Tsang, T., Coyle, G. J., Adler, I., and Yin, Y. (1979) XPS studies of ion bombardment damage of iron–sulfur compounds. *J. Electron Spectron Spectrosc.* **16**, 389–396.
88. Kim, K. S. and Winograd, N. W. (1974) X-ray photoelectron spectroscopic studies of nickel–oxygen surfaces using oxygen and argon ion bombardment. *Surface Sci.* **43**, 625–643.

89. Kim, K. S., Batinger, W. E., Amy, J. W., and Winograd, N. W. (1974) ESCA studies of metal–oxygen surfaces using argon and oxygen ion bombardment. *J. Electron Spectrosc. Rel. Phen.* **5**, 351–369.
90. Yin, L., Tsang, T., and Adler, I. (1975) Electron spectroscopic studies related to solar wind darkening of lunar surface. *Geophys. Res. Lett.* **2**, 33–36.
91. Yin, L., Tsang, T. and Adler, I. (1975) ESCA studies on solar-wind reduction mechanism. *Proc. Lunar Sci. Conf. 6*, **3**, 3277–3285.
92. Yin, L., Tsang, T., and Adler, I. (1976) On the ion-bombardment reduction mechanism. *Proc. Lunar Sci. Conf.* 7, **1**, 891–900.
93. Storp, S. and Holm, R. (1979) ESCA investigations of ion beam effects on surfaces. *J. Electron Spectrosc. Rel. Phen.* **16**, 183–194.
94. Chuang, T. J., Brundle, C. R. and Wandelt, K. (1979) An XPS study of the chemical changes in oxide and hydroxide surface induced by Ar^+ ion bombardment. *J. Vac. Sci. Technol.* **16**, 797.
95. Katrib, A. (1980) The reduction of Pt(IV) to Pt(II) by X-ray and argon-ion bombardment; evidence from X-ray photoelectron spectroscopy. *J. Electron Spectrosc. Rel. Phen.* **18**, 275–278.
96. Jacobi, K. and Ranke, W. (1976) Oxidation and annealing of GaP and GaAs(III) faces studied by AES and UPS. *J. Electron Spectrosc. Rel. Phen.* **8**, 225–238.
97. Coyl, G. J., Tsang, T., Adler, I., and Yin, L., (1981) XPS studies of surface damage of transition metal fluorosilicates under argon ion bombardment. *Surf. Sci.* **112**, 197–205.
98. Zilinskas, E., Skorobogatas, H. *et al.* (1983) Temperature dependence of compositional changes of SiO_2 surface during ion and electron bombardment. *Surf. Sci.* **134**, 464–468.
99. Kraševec, V. and Navinšek, B. (1974) Surface oxidation of ion bombarded NiMn alloy. *Surf. Sci.* **45**, 39–44.
100. Christie, A. B., Lee, J., Sutherland, I. *et al.* (1983) An XPS study of ion-induced compositional changes with group II and IV compounds. *Appl. Surf. Sci.* **15**, 224–237.
101. Taylor, J. A., Lancaster, G. M. and Rabalais, J. W. (1978) Interaction of N_2^+ and NO^+ ions with surfaces of graphite, diamond, teflon and graphite monofluoride. *J. Am. Chem. Soc.* **100**, 4441–4447.
102. Taylor, J. A., Lancaster, G. M., and Rabalais, J. W. (1978) Chemical reactions of N_2^+ ion beams with group IV elements and their oxides. *J. Electron Spectrosc. Rel. Phen.* **13**, 435–444.

Chapter 4

Study of adsorption and catalysts

4.1. X-ray photoelectron study of the adsorption of molecules on metals

At present there exists extensive literature (see references in reviews [1–5]) devoted to the X-ray-photoelectron study of adsorption. This field of applied electron spectroscopy is one of the most complex both in terms of staging experiments and of interpreting their results. The difficulties in experiments are connected with the need to work in conditions of super-high vacuum and, most important, the need to obtain pure metal surfaces. The presence of traces of oxygen can noticeably influence the results of studies. The interpretation of results is impaired by the formation of several products of adsorption (with the resolving power of the spectrometer often turning out to be insufficient for separating their spectral characteristics) and by the inadequate contrast of spectra. In addition, there exist fundamental difficulties connected with the choice of the reference zero to determine kinetic energies (see further and [5–10]), with the absence of simple regularities between the change of the spectral characteristics of adsorbed molecules and the parameters determining the process of adsorption. As a result of these factors, experimental data and their interpretation by various authors often vary greatly. As a whole, this field of study is characterized by flux—concepts concerning the interpretation of data quickly change, and the number of experimental results grows daily. All this renders a detailed presentation of the vast body of existing material impractical, and for this reason we shall limit ourselves to a presentation of general results and a study of the main tasks facing researchers: identification of products of adsorption and the interconnection of spectral characteristics and the process of adsorption.

4.1.1. *Identification of products of adsorption*

The adsorption of CO by metals can lead to various types of bonding between CO and the metal molecule. As a rule [1, 2, 11] three types of bonding are discerned, α, β, and υ (virgin; Fig. 4.1), where the nature of CO–metal bonding can be viewed as substantiated only in the case of the α-state. The nature of bonding for β- and υ-states is conjectural. There exist, apparently, many forms of the β type that are characterized by a differing degrees CO bonding, right up to dissociation into O and C atoms. It is thought that υ state develops into β bonding when the temperature is increased. The identification of these states on the basis of spectral data is as follows.

α-type states (they are sometimes designated γ) comply with O 1s energies

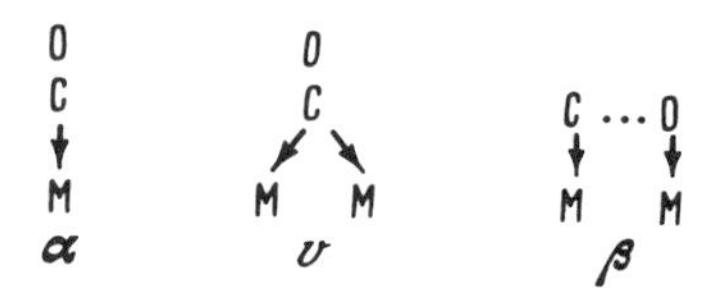

Figure 4.1. Various bonding forms (α, β, υ) and adsorption of CO in metals.

around 531.7–537 eV and C 1s energies of 285–290 eV. υ-states are characterized by energies C 1s = 285 eV and O 1s = 531 eV. The spectra of valence levels of adsorbed α-type CO molecules are analogous to those of a gaseous CO molecule.

The state of type β accords with O 1s energies near 530 eV and C 1s around 283 eV. Lower values of C 1s and O 1s indicate the formation of metal–oxygen and metal–carbon bonds close to those in oxides and carbides. Only a broad minimum of 6.6–8 eV from the Fermi level is observed in the spectra of valence levels. Data for various states of CO in Mo/CO and W/CO systems are cited in Table 4.1 [1]. It should be noted that α- and υ-states were observed during the adsorption of CO on Ni, Pt, Pd, Ir, Cu, Fe, Ru, Re, W, and Mo, while α-states were seen during the adsorption of CO molecules on W, Mo, Fe, Ni, and Ti [2, 3, 12]. According to [13] the formation of β_2-states is characterized by a large CO adsorption energy ($H_{ads} > 250\,\mathrm{kJ\,mol^{-1}}$ at 290° C). It is shown in the same study that the value of O 1s for adsorbed CO molecules decreases with the growth of H_{ads}. This correlation testifies to the π-mechanism of bonding between the molecule of CO and the atom of metal: the transfer of electron density from d-orbitals of metal to the π-orbital of CO leads to a lessening of O 1s energy and a growth of H_{ads}.

It must be noted that great caution should be displayed when comparing various maxima of O 1s and C 1s with various states of CO on the surface of metal (although at present this is rather widespread). Indeed, intensive satellites of C 1s and O 1s lines in carbonyls are known and it has been shown that intensive satellites for adsorbed CO molecules can be connected with a number of many-electron processes [14].

Numerous X-ray photoelectron studies of CO and K coadsorption on various metals (see literature in [15–17]) have appeared lately in connection with the large influence of additions of alkaline metals on the selectivity and catalytic activity of catalysts.

The adsorption of NO by metals also leads to the formation of a number of

Table 4.1
C 1s and O 1s (eV) binding energies

State	C 1s	O 1s	C 1s	O 1s
	Mo/CO		W/CO	
β	282.7	530.0	282.6	529.9
$\gamma(\alpha)$	288.0	534.0	287.0	533.0
υ	284.6	531.2	284.8	530.9

Table 4.2
Data on the Ni/NO system

Method of treatment		N 1s, eV	O 1s, eV	(Ni:O:N), atomic
(I)	Adsorption of NO at 77 K	399.9	530.9	1:0.46:0.40
(II)	Heating of (I) to 300 K	397.8; 399.9	529.5; 530.9	1:0.46:0.42
(III)	Addition of NO to (II) at 300 K	397.8; 403.0	529.5; 531.7	1:1.61:0.66

products. Results of [18] for the Ni/NO system are given in Table 4.2. Data concerning N 1s and O 1s during adsorption at 77 K are interpreted as a result of the presence of the molecular adsorbed form NO. Heating to 300 K leads to a partial dissociation of NO and the forming of M—N and M—O bonds that are close by type to bonds in a nitride and an oxide (see N 1s lines at 391.8 eV and O 1s at 529.5 eV, Table 4.2). When NO is added to this system at 300 K this results in a dissociation of NO and loss of nitrogen (see O and N ratio in Table 4.2). Along with the nitride and oxide fractions there are also on the surface weakly bound NO molecules associated with the energies N 1s = 403 eV and O 1s = 531.7 eV.

Somewhat different results for the same Ni/NO system were obtained in [19]. Higher resolution made it possible to detect an additional structure in the N 1s line (Fig. 4.2). Present in spectra of various intensity are N 1s maxima with the energies: (1) 397.5; (2) 399.5; (3) 402.5; and (4) 405.1 eV. They are referred: (1) to the nitride Ni—N; (2) bent Ni—NO; (3) linear Ni—NO forms and to N_2O; and (4) to N_2O. In accordance with this interpretation adsorption at 80 K is accompanied by forms Ni—N, Ni—No and N_2O (see Fig. 4.2, spectrum 4). Adsorption of NO on the surface Ni oxidized at 80 K leads to Ni—NO bent and

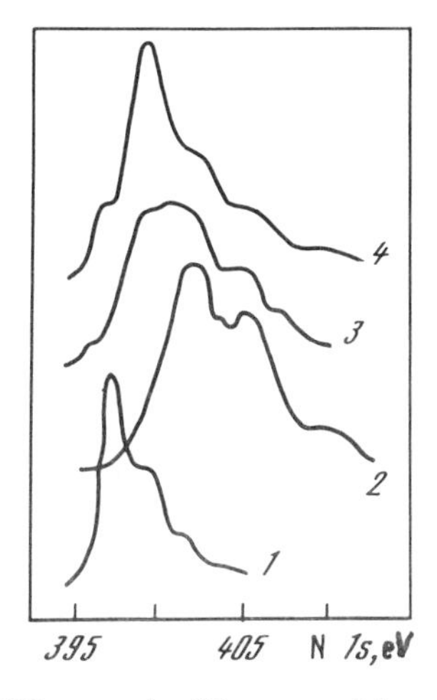

Figure 4.2. N 1s line of the Ni/NO system in different conditions of adsorption and subsequent treatment.

linear forms and N_2O (spectrum 3). During the adsorption of NO on the surface Ni oxidized at 290 K we obtain mostly linear Ni—NO form and N_2O (spectrum 2). When adsorbed on metallic Ni at 290 K the main mass of No dissociates (spectrum 1). Data for the systems Cu/NO, Fe/NO, and Al/NO are analogous to those given above for the system Ni/NO [19]. Bridge and linear groups of NO [20] have also been discovered in the system Pt(111)/NO. Data on the formation of linear and bridge-adsorbed NO molecules in the system Ru(001)/NO at temperatures lower than 200 K are presented in [21]. Dissociation of NO is observed at temperatures above 200 K. Adsorbed NO molecules withdraw at 450 K, followed by all nitrogen atoms at 600 K, leaving only atoms of oxygen on the surface. The same Ru(001)/NO system is studied in [22].

The dissociation of NO occurs more easily on Ru, Ir, Ni, and Fe, and less actively on W and Pt (see literature in [3]). It is interesting to note that the dissociation of NO on the oxidized surface of nickel is noticeably weakened because the maximum intensity at 397.5 eV (Fig. 4.2) is not great, i.e. oxygen blocks the active centres, facilitating the dissociation of NO. An analogous result has been obtained for the system Ir/NO [23]. The Co/NO system is studied in detail in [24].

In principle, the interpretation of the N 1s spectrum of an adsorbed NO molecule can be impaired by the presence of an unpaired electron on the π^*-orbital of NO. In that case the N 1s line can be split even if there is only one type of NO bonding with the metal. Usually this possibility is not discussed in the literature, but it is noted in [25] that the splitting of the N 1s maximum in the Ir/NO system at 170 K can be explained exactly by this mechanism.

The adsorption of O_2 by metals can also bring about the formation of several products. It has been noted already [26] that the O 1s energy for the first stages of adsorption is approximately constant for all metals (although use was made of data obtained by various authors using different energy graduations). Nevertheless, according to Brundle [27], there really is a maximum for O 1s, with an energy of 530 ± 0.5 eV, for the majority of metals that had been studied by then (Ni, Fe, Na, Ag, V, Mo, W, Zn, Ru, Ir, Au, Mn, Cr, Co). This maximum is imputed to oxygen that has formed a bond with metal similar to the metal–oxygen bonding in oxides. Indeed, this value of energy O 1s is close to that in the corresponding oxides (see Table 1.2). According to [5, 28] the maximum O 1s with an energy of about 532–533 eV, which can be found in many oxides, refers to OH groups. A big selection of O 1s values in oxides, hydro-oxides and frozen ice depending on the metal substrate is given in [29]. See also review [30].

Along with the main maximum close to 530 eV, other maxima of O 1s have also been discovered for a number of metals, for instance Zn, Co, and Ni during the adsorption of O_2. Thus [31], during the adsorption of Zn/O_2 there was observed an O 1s maximum with an energy of 531.8 eV, which was referred to the chemisorbed oxygen atoms in the uppermost part of the metal's surface, or in ZnO, or between Zn and ZnO. As a result, these oxygen atoms differ from other atoms of oxygen in forming a bond close to that in ZnO (O 1s ~ 530.0 eV).

There are three O 1s maxima with energies of 529.5, 531, and 533 eV in the spectrum of the Co/O_2 system [32]. The first maximum is associated with an oxide, the second with OH^- groups, and the latter with H_2O. The O 1s spectrum in the Cu/O_2 system is interpreted similarly. Three peaks (O 1s with energies of 528.3, 530.3, and 532.5 eV) differing by about 1 eV from the value in the Co/O_2 system have also been found in the Ag/O_2 system [33]. But a different interpretation is offered: the maxima relate to chemisorbed oxygen, oxygen in volume, and oxygen on the surface. The same systems are studied in [34].

The anions O^{2-}, O_2^{2-}, and O_2^- have been discovered on the surface of oxidized Cs [35]. A review of studies dealing with the Al/O_2 system is presented in [36].

O 1s maxima with energies of 530.2, 535.6, and 533.4 eV were discovered during the adsorption of O_2 on Re(0001). The first maximum was referred to the atomic state, the second to the molecular state (stable at temperatures from 50 to 80 K), and the third to the transition state (stable from 80 to 300 K). Molecular states during the adsorption of O_2 at low temperatures are quite usual (see references in [37]).

A particularly complex picture is presented by the O 1s spectrum in the Ni/O_2 system [1, 31, 38, 39] (Fig. 4.3 [38]). The following interpretation is suggested, taking account of the values of energies and intensities of individual maxima, depending on the treatment of nickel. Maximum I (529.5 eV [33] is referred to atoms of oxygen that have formed bonding with nickel similar to that in oxide or to chemisorbed oxygen atoms. Maximum II (531.4 eV) is referred [38] to atoms of

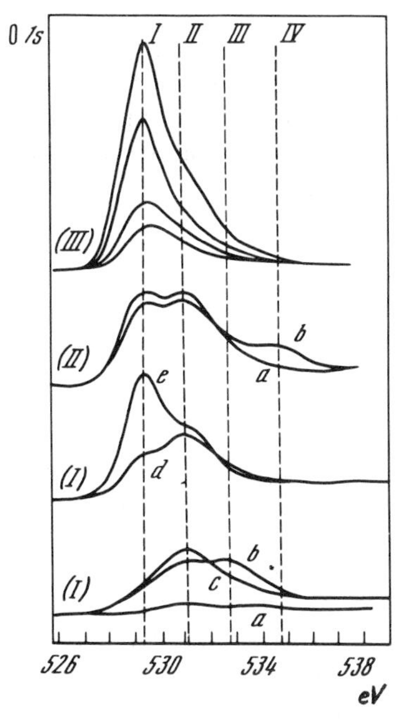

Figure 4.3. O 1 s line of Ni/O_2 system. (I) a—before the commencement of oxygen adsorption; b—saturated adsorption of 10^{-5} torr O_2 at 77 K; c—heating b to 300 K; d—further adsorption of O_2 in the course of 2 min at 2×10^{-7} torr and 300 K; e—continued adsorption of O_2 at 300 K. (II) a and b—see (I) a and (II) b for other Ni film. (III) Growing adsorption of O_2 up to saturation of Ni at 400 K.

oxygen on the nickel surface that have formed a weak bonding with the nickel (surface chemisorption). Later, this maximum was interpreted as a result of the presence of OH^- groups at least for adsorption at 300 K [33]) or as the spectrum of an O atom in a defective Ni_2O_3 oxide [40]. Maximum III (533.2 eV) is referred to the initial stage of oxygen chemisorption that is stable at low temperatures (perhaps, Ni—O_2) and passes into state II when heated. The presence of a fourth O ls maximum with an energy of about 535 eV [33] is not ruled out.

It should be noted that the Ni $2p_{3/2}$ line (Fig. 4.4) does not change in practice (compare (1) a, b, c, d, e in Figs 4.3 and 4.4) under various conditions of the reaction of Ni with O_2, despite the noticeable change in the O 1s profile. A noticeable change of Ni $2p_{3/2}$, testifying to the beginning of the forming of an oxide on the surface, is observed only in the event of a reaction of Ni with O_2 at a high temperature (400 K). It should be noted that the absence of changes in the Ni line at the initial stage of the adsorption of oxygen and other molecules is quite typical. It is explained by two circumstances. First, the contribution of the atoms of the metal surface interacting with the adsorbed molecules to the overall intensity of the line is comparatively small. Second, the adsorbed molecule's interaction with the metal is largely of a 'collective' nature in the sense that the molecule interacts with the metal's entire electron system and not with an individual atom, and for this reason the perturbation of the metal's electron state is comparatively small. This 'collective' nature of interaction is confirmed also by data on changes in the vibration frequency of adsorbed CO molecules. During the adsorption of CO a part of the metal's electron density passes to the antibonding π-level of CO and the v (CO) vibration frequency decreases as compared to the free molecule. This decrease in Δv is often greater than in carbonyls, where there are one or more CO groups per atom of metal. With the growth of the number of adsorbed CO molecules on the surface of metal,

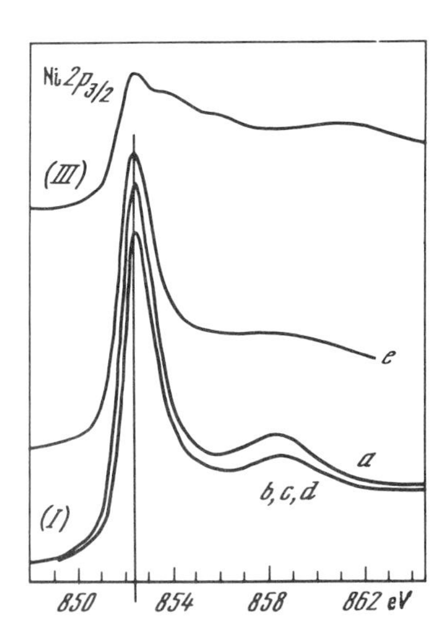

Figure 4.4. Ni $2p_{3/2}$ line in Ni/O_2 system. (I) and (III)—see Figure (4.3).

the competition of the CO molecule for the metal's electron density increases and the value of $\Delta \upsilon$ decreases.

In principle it is possible to increase the sensitivity of the method to the topmost surface layer by changing the photoelectron exit angle. In a number of instances it has proved possible to detect small changes in the bonding energies of the substrate atoms, depending on the extent of covering (for instance, Ru $3d_{5/2}$ grows by 0.1–0.3 eV in Ru/O_2 and Ru/CO systems with a growth of coverage rate θ from 0 to 1 [41]). For the O_2/Ru(001) system the higher the surface sensitivity (the photoelectron exit angle to the surface is less) the greater the positive shift in the metal lines. The shift monotonously grows with the coverage rate of O_2. In contrast, [42] shows not a monotonous shift of W4f line with the growth of surface sensitivity, but the appearance of an additional maximum with an energy greater by 0.9 eV than the main one. This data points to noticeable differences in the structure of the metal–oxygen surface layer in various systems.

A change in the nature of the Pt valence band (decrease in the density states near the Fermi level) after the adsorption of O_2, CO, and H_2O was discovered in [43]. It was shown in [44], that in a partially reduced $SrTiO_3$ (111) monocrystal containing Ti (III), a decrease in the number of Ti (III) atoms is observed after adsorption of O_2, H_2, and H_2O.

Above we studied examples of determining various adsorption states using the spectra of inner levels. This can also be accomplished on the basis of valence level spectra, in particular for CO molecules. Admittedly, this procedure is extremely complex and is connected with a number of assumptions and often with the need for a theoretical evaluation of the position of levels [see, for instance, [45] on the identification of adsorption products of C_2H_2 on Ni(100) and (110)].

Studies of recent years also indicate the dependence of the nature of adsorption and subsequent reactions of the adsorbent on the type of monocrystal surface. In [46], two forms of adsorbed NO molecule were found on the surface of iridium (111), while only one was found on the surface of iridium (110). Moreover, the latter form dissociates at much higher temperatures than the form on the surface of iridium (110). It was found in [40] that the speed at which the oxide grows on the planes of a nickel monocrystal in the temperature range 295–485 K changes in the order (110) > (111) > (100). Both the valence and inner level Al 2p point to the differential adsorption of oxygen by planes (100) and (110) of an aluminium monocrystal [47].

It follows from the above that both dissociative and molecular mechanisms of adsorption are observed during the adsorption of CO, NO, and O_2. This is true also for the adsorption of N_2 [48]. An interesting correlation is noted in reference [48]. Dissociation energy decreases in the order CO, N_2, NO, and O_2. In accordance with this result, dissociative adsorption at room temperature for the O_2 molecule occurs for virtually all metals, with the exception perhaps of platinum. The ability of other metals in terms of dissociative adsorption of CO, N_2 and NO at room temperature depends on their position in the Periodic Table (Fig. 4.5): the adsorbed molecule's dissociation capability decreases with decreasing metal electronegativity. The NO molecule being more inclined to dissociation than CO and N_2. These correlations point to the interconnection of the nature of

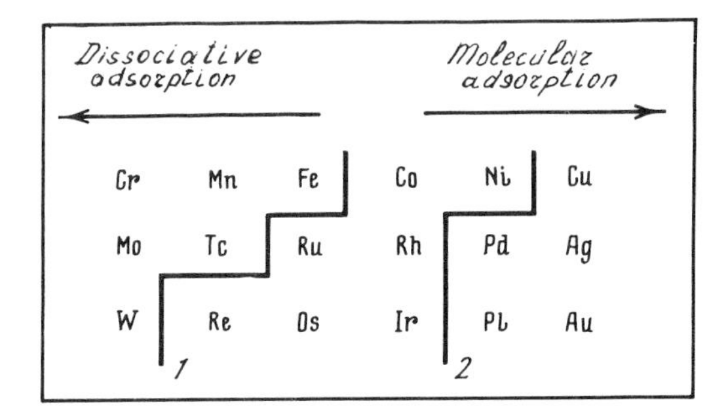

Figure 4.5. Nature of adsorption. 1 = border for N_2 and CO; 2 = border for NO.

adsorption and the amount of electron density transferred from the substrate to the adsorbate. Of course, these results should not be taken as absolute, especially in borderline cases. Thus molecular adsorption is observed during the adsorption of Re(0001)/CO, while both molecular and dissociative adsorption are observed in the case of Re (polycrystal)/CO [49]. The plane Pt(410) [50] displays an unusually high aptitude for dissociative adsorption of NO and CO.

4.1.2. *Problems of selecting the energy reference level*

Late in the 1970s there appeared a number of studies [1, 5–10] devoted to the energy reference levels, especially for the adsorbate–substrate systems. The stated viewpoints are contradictory. In [7], for instance, it is suggested that the substrate energy level should be regarded as a reasonable reference for determination of kinetic energies because the Fermi levels have a permanent value and are a sort of universal value (the Earth's Fermi level for grounded samples). In [8] it is contended that to obtain absolute binding energies for substrate and adsorbate electrons it is enough to add the spectrometer's work function to the binding energies defined with respect to the spectrometer's Fermi level. Both studies fail to take into account the change in the state of the sample under investigation (actually the charging) as a result of contact with the spectrometer material. In [6] it is proposed to determine the absolute binding energies in the adsorbate by adding the work function of a surface covered with a monolayer of the adsorbate to the measured binding energies. Further information on the essence of the problem will be presented on the basis of notions outlined in [9, 10].

It should be noted first of all that, by definition, the electron binding energy with respect to vacuum level should always be considered (see Chapter 1) equal to

$$E_b^v = E_b^F + \Phi_s, \tag{4.1}$$

where Φ_s is the work function of the system under study, and E_b^F is the binding energy relative to the Fermi level. The essence of the problem is to equally compare the binding energy of an adsorbed atom or atom in a solid with that of a free atom. The problem is that a formally calculated shift ΔE_b of an atom in a solid

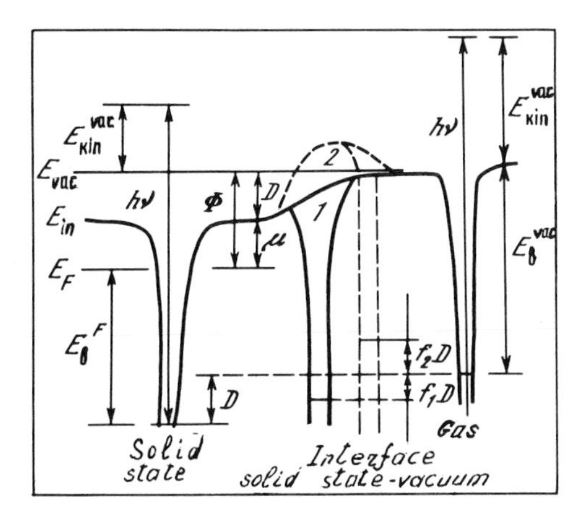

Figure 4.6. Energy level diagram. E_{vac} = vacuum level; E_{in} = average potential inside solid; E_F = Fermi level. 1, 2 = possible transitions from E_{in} to E_{vac}.

as compared to a free atom equals

$$\Delta E_b = E_b(\text{ads}) - E_b(\text{gas}) = \Delta E_{ads} + \Phi_s, \tag{4.2}$$

where $E_{ads} = E_b^F(\text{ads}) - E^v(\text{gas})$. This shift contains a part connected with the change in the photoelectron kinetic energy when passing the solid–vacuum border, this change in kinetic energy only partly reflecting the difference between the atom in the solid and the free atom. A substantial contribution to this change is made by the surface dipole potential, which reflects the state of the surface and depends on its purity, the type of facet of the monocrystal plane, etc. Work function Φ can be presented in the form of two parts (Fig. 4.6, $\Phi > 0$, $D < 0$, $\breve{\mu} < 0$*); $\Phi = -D - \breve{\mu}$, where $\breve{\mu}$ is the chemical potential, i.e. the average potential inside the solid E_{in} with respect to the Fermi surface.

If we are to study the case when the atom is in a solid (metal) and does not interact with it, the kinetic energy of this atom's photoelectron will all the same change by the value D, i.e. $E_b(\text{metal}) = E_b(\text{gas}) - D$. If the adsorbed atom or molecule is over the surface or on the surface without interacting with it, account should be taken of only a part of the dipole potential D, and in that case we have

$$E_b(\text{ads}) = E_b(\text{gas}) - fD, \tag{4.3}$$

with $0 < f < 1$.

Consequently, two extreme instances can be considered. If the adsorbed particles are outside the surface dipole ($f = 0$) of the substrate, the measured $E_b^F(\text{ads})$ values will depend on the work function of the studied sample Φ_s and change with it, for instance, with an increase of the extent of the surface's coverage. In the extreme instance of a thick adsorbate, the value Φ_s is equal to the

*The choice of signs shows whether the changes in the photoelectron kinetic energy as a result of the values Φ, D and $\breve{\mu}$ during the solid–vacuum transition accord with the sign of θ, D, and $\breve{\mu}$ or not. Binding energies are regarded as positive values.

adsorbate's work function. If $f = 1$, i.e. the particles are in the substrate, then the change in the work function of the studied surface of the substrate should not influence changes in the energy E_b^F(metal) in the event of thin adsorbate coverage of the substrate surface. It is obvious, however, that in the event of substantial adsorption such changes should take place [transition from Φ(sub) to Φ(ads) in (4.2)].

This approach [9] formulates the problem but does not provide any concrete recommendations on how to determine fD in (4.3). It appears difficult at present to offer a clear-cut solution of the problem. In reference [10], containing a detailed analysis of this problem, it is proposed in (4.2) to replace Φ_s, the work function of the surface being studied (which depends on the extent of coverage), with Φ, the work function of the pure substrate. To avoid misunderstandings we emphasize that the point is not to determine the value E_b^v(ads) (in that case it would be necessary to add Φ_s), but to obtain a certain definite value in which a member of the fD type is ignored.

It should be noted that a change in the value Φ for various planes of the monocrystal usually noticeably influences the values E_b, with ΔE_{av}^F usually being close to the difference $\Delta\Phi$ for various planes. For example, during the adsorption of CO on the planes W(100) and (110), the difference of the inner energy levels of CO amounts to 1 eV, which is close to the difference of the work functions of these planes [51].

4.1.3. *Shifts of inner and valence levels during adsorption*

A change in the electron density of the adsorbate takes place during adsorption as a result of interaction of adsorbate and substrate. This change should be reflected in the values of the binding energy for the inner and valence levels. As has been explained above (see Chapter 1), the shift of the inner level ΔE can be presented in the form

$$\Delta E = k\Delta q + \Delta V + \Delta E_{rel}.$$

The first two components are connected with the initial (non-ionized) state of the molecule, while the latter reflects the redistribution of electron density in the molecule after photoionization. As a rule, for a given atom in different solids the value ΔE_{rel} is smaller than Δq and ΔV and for this reason ΔE reflects the change in the molecule's initial state and usually correlates with Δq. But the value ΔE_{rel} is substantial when studying the value ΔE for the gas–solid or free molecule–adsorbed molecule transition because during such transition the possibility of drawing away electron density to the ionized atom from other atoms or molecules in the solid (extra-atomic relaxation) sharply increases.

Indeed, during the physical adsorption of Xe on W(111) the work function drops by 1 eV. This is interpreted by the authors [52] as a result of the transition of a part of the Xe charge to W, which should cause the positive shift Xe $3d_{5/2}$ after adsorption. But what is observed is a negative shift of Xe $3d_{5/2}$ by -3.2 eV, thus testifying to the dominant role of the ΔE_{rel} member in the expression for ΔE.

The results obtained for Xe are typical: for the transition from free molecules to

Table 4.3
Energy level shift (eV) as compared to free molecules

Molecule	Line	ΔE_{in}	ΔE_{val}	Molecule	Line	ΔE_{in}	ΔE_{val}
H_2O (cond)	O 1s	6.6	~5.5	CO_2 (ads)	O 1s	6.9	
H_2S (cond)	S 2p	6.2	6.4	CO_2 (cond)	C 1s	5.7	5.3
H_2S (ads)	S 2p	7.7	6.2		O 1s	5.8	
	C 1s	6.6	6.5	SO_2 (cond)	S 2p	6.9	7.0
					O 1s	6.8	

a solid (condensation or adsorption) the value ΔE is less than zero. This was observed, for instance, for valence levels of condensed H_2CO, H_2O, C_6H_6, Py, CH_3OH, NH_3, and C_2H_5OH on MoS_2 [53], and for valence and inner levels H_2O, H_2S, CO, and SO_2 after condensation of Ni and Cu at 80 K [54] (see Table 4.3).

When analysing Table 4.3 it is necessary to bear in mind that the energy levels for the solid were determined in the study [54] with respect to the Fermi level, and for free molecules with respect to the vacuum level. For this reason the value ΔE is in reality smaller than the values cited by the amount of the work function (roughly by 5 eV). It follows from Table 4.3 that the shifts due to adsorption and condensation are approximately equal (in the case of adsorption they are greater) and that the shifts of inner and valence levels in the molecule often coincide.

Let us note the following patterns in the shifts of the free molecule valence levels due to condensation and adsorption [1, 53, 55, 56]. After condensation all valence levels shift by a constant value (towards lesser energies). After adsorption all valence levels that do not take part in interaction with the substrate shift by a constant value. The molecular levels are either electron density donors with respect to the substrate (for example the 5σ level in CO) or acceptors (for instance the π-levels in C_6H_6, C_2H_2, and C_2H_4, increase their ionization energies as compared to other levels of the molecule). As illustrations, the spectra of C_6H_6 are shown in Fig. (4.7) [1], diagrams of C_6H_6, C_2H_2 and C_2H_4 are shown in Fig. (4.8) [55], and a diagram of CO levels in Fig. (4.9) [56]. A review of data on NO adsorption is given in [57].

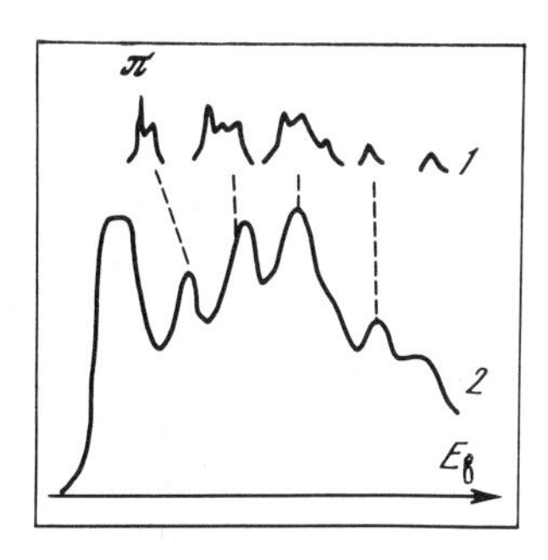

Figure 4.7. He(I) spectra for C_6H_6. 1 = free molecule; 2 = molecule adsorbed at 77 K on Ni.

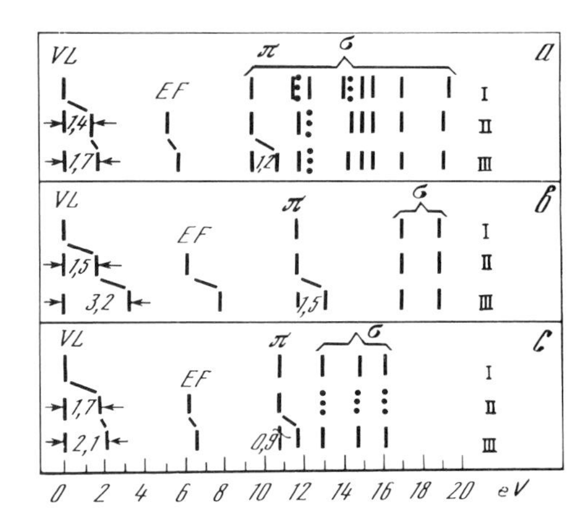

Figure 4.8. Diagram of (a) C_6H_6, (b) C_2H_2, and (c) C_2H_4 valence levels. I = free molecule; II = condensate on Ni(III) at 100 K; III = chemisorption on Ni(III) at 300 K. VL = vacuum level, E_F = Fermi energy.

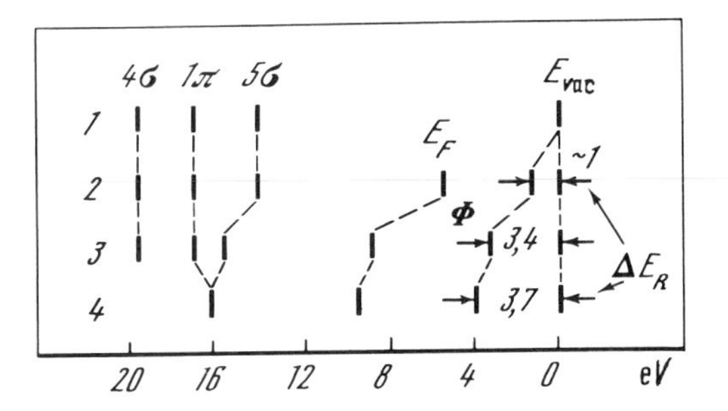

Figure 4.9. Diagram of CO valence levels aligned by the 4σ level. 1 = free molecule; 2 = physical adsorption of 20 K on Cu; 3 = weak chemisorption at 98 K on Fe(100) + S; 4 = strong chemisorption on Fe(100) at 123 K. E_F = Fermi level, Φ = work function, ΔE_R = relaxation shift, E_{vac} = vacuum level.

When studying Figs (4.7–4.9) one sees, first, that the observed patterns of change in the energies of the molecule's valence levels due to adsorption are fully analogous to those that occur during the coordination of the same molecule by a transition metal to the complex (see, for instance, data on CO and C_6H_6 molecules [58, 59]. Second, the extent of the shift in the levels of molecules taking part in the chemical bond with the substrate depends on the 'strength' of this bond (Fig. 4.9). Third, although the energy distance between most levels after adsorption does not change small changes still take place (within the limits 0.1–0.3 eV) which can be used for studying the specificities of the molecule's adsorption on the given substrate. For example, the 1π–4σ distance in free CO equals 2.8 eV, but after adsorption the value often increases by several several tenths of an electron volt [48]. This increase can be linked with the greater weakening of the π-bond as compared to the σ-bond during adsorption as a result of a small increase of the C—O distance [48]. This change is particularly great for the adsorption of CO on Pt [60], this being evidence of strong chemisorption.

Last but not least, one should take into consideration the possibility of the appearance of new valence levels during adsorption [61].

In reference [55] it was proposed to regard the value ΔE_{rel} as constant for all levels of the adsorbed molecule. Since in practical terms for most levels participation in the adsorbate–substrate interaction should be equal to zero (for instance, for the levels 4σ and 1π in CO), ΔE_{rel} for these levels coincides with ΔE. This makes it possible to calculate the ΔE share for π-levels in C_6H_6, C_2H_4, C_2H_2 and for the 5σ level in CO, which is caused by the interaction of these molecules with the substrate (ΔE_{ads}). The values ΔE_{rel} and ΔE_{ads} are given in Figs 4.8 and 4.9. The value ΔE_{ads} is smaller than ΔE_{rel} and amounts to about 1 eV for π-levels and 2 eV for the 5σ-level of CO.

The correlation of the value ΔE_{ads} with the adsorption energy C_2H_2 and C_2H_4 on Ni is shown in [55]. It is seen from the more detailed study [62] of the adsorption of C_2H_2 and C_2H_4 on Fe, Ni, and Cu that there is no correlation of the value ΔE_{ads} either with the experimental value of the adsorption energy or with the calculated energy of the formation of the molecular complex with the substrate. The authors of the study [62] believe that to take account of bonding strength it is not enough to consider only the change of ΔE_{ads} in the adsorbed molecule but that it is also necessary to take into account changes in the substrate. One can accept this but the very nature of the calculation of ΔE_{rel} cannot be regarded as sufficiently substantiated because ΔE_{rel} for the adsorbate–substrate interaction levels should differ from ΔE_{rel} for 'passive' levels. Indirectly this dependence follows already from the fact that the value of ΔE_{rel} for 'passive' levels depends on the matrix where the molecule under study is situated. Thus, the ΔE_{rel} value of 'passive' CO and N_2 levels in solid CO and N_2 and in an Xe matrix was determined in [63]. It turned out that the value ΔE_{rel} grows in the series N_2 in N_2 (0.8 eV), CO in CO (1.0 eV) and N_2 and CO in Xe (1.4 eV), this reflecting the growth of matrix polarization.

Binding energies both of inner and valence levels, measured with respect to the substrate's Fermi level, depend on the properties of the substrate (Table 4.4) [64]. These differences remain after the corresponding work function has been added. But in the case of physical adsorption ($T = 20$ K) of Ar, Kr, Xe, CO, O_2, and N_2 gases on Ni(110) at low temperatures, differences of up to 3 eV in the binding energies of the adsorbed molecules (depending on the extent of the substrate coverage) are explained by changes in the substrate's work functions. A constant value of binding energy that does not depend on the extent of substrate coverage

Table 4.4
Influence of substrate on CO (eV) energy levels

Level	CO (gas)	W(110)	Ni(100)	Ni(111)	Pt	Cu
1σ (O 1s)	542.6	531.6	531.6	531.0	532.7	532.8
2σ (C 1s)	296.2	285.5	285.5	285.9	286.6	286.3
3σ	38.3		29.0	29.0		
4σ	19.7	10.5	11.2	11.2	11.7	11.9
1π	16.8	7.2	8.2	8.0	9.1	8.8
5σ	14.0	8.0	7.0	7.0	8.0	8.0

and the nature of the substrate is obtained when the binding energy is referred to the vacuum level E_b, i.e. $E_b = E_b^F + \varphi = \text{const.}$ [65].

This circumstance is used in [66] to propose a new method of studying the local work function φ_L for heterogenous surfaces by reversing the equations cited above: $\varphi_L = \text{const.} - E_b^F$ on the basis of measured values E_b^F. This method makes it possible, for instance, to study defects on the surface, differences in the activity of the alloy's various atoms during adsorption depending on the coverage rate. It opens up the very important possibility of 'titrating' surface defects or studying surface structure [67].

It is proved in [68] that the energies of CO valence levels adsorbed on the Cu–Ni alloy change depending on the atom that coordinates the CO molecule. A difference (~ 0.6 eV) between the energy of CO adsorbed on a copper atom in the Cu–Ni alloy and CO adsorbed on metallic copper has also been found. Analogous effects have been studied for the adsorption of CO for various planes of a monocrystal of the alloy $Pt_{0.98}Cu_{0.02}$ [69].

X-ray photoelectron data reveal, in addition, the dependence of the adsorbate binding energy on the coverage rate of the adsorbate surface. This means that account should be taken of the interaction of chemiadsorbed atoms and molecules in which the adatom–adatom interaction can occur through the substrate [70].

Let us dwell on some theoretical papers studying the interconnection of the values ΔE, ΔE_{ads}, and ΔE_{rel} (see review [71]). The values ΔE, ΔE_{ads}, and ΔE_{rel} in NiCO and Ni_2CO clusters were calculated by the Hartree–Fock method in [72]. Agreement between calculated and experimental values was not satisfactory: the computations did not give constant ΔE values for 1π and 4δ levels in the CO molecule. Moreover, and the sign of ΔE was computed erroneously. The authors explain the latter error by underestimation of the value ΔE_{rel}. The $CONi_5$ cluster was estimated by the $X\alpha$ method in [73]. In this paper the energies of CO valence levels accorded well with the experimental ones but the values ΔE_{ads} and ΔE_{rel} were not determined.

Of special interest are the results of study [74], in which Al_5X octahedral clusters were calculated, with X = Na, Si, Cl. The quantity of the charge q that goes over into the sphere of atom X as a result of extra-atomic relaxation after photoionization was determined. The value $q = 0.5, 1.1, 0.8$ e for X = Na, Si, and Cl, respectively. The distribution of this additional electron density on cluster levels after the photoionization of atom X was also estimated. It turned out that the share of the charge of atom X on various electron levels changes noticeably as a result of photoionization. Moreover it is easy to explain these changes as a result of the increase in the X atom's orbital energies after photoionization. For instance, (Table 4.5), as a result of the increase in the Cl 3p orbital energy after photoionization, the population of this orbital has noticeably increased for the more low-lying levels $2a_1$ and 1 e, while decreasing for the higher $3a_1$, $4a_1$ and 2 e levels. These population changes are noticeably greater than the expected changes after X adsorption on Al_5, thus testifying to the importance of the contributions of ΔE_{rel} and ΔE. It is also seen from the results obtained that the $\Delta E_{rel} = \text{const.}$ equation for 'active' and 'passive' levels should not be too strict.

Table 4.5
Population distribution of Cl orbitals in Al_5Cl (A) before and (B) after photoionization

Level	A	B	Level	A	B
$1a_1$	1.872	1.929	$3a_1$	0.986	0.365
$2a_1$	0.214	1.410	2e	1.517	0.168
1e	1.598	3.420	$4a_1$	0.396	0.058

Reference [64] offers a theoretical evaluation of the contribution by ΔE_{ads} and ΔE_{rel} to ΔE for a number of atoms adsorbed on a metal-imitating substrate. It is also shown in this paper that for an adsorbed CO molecule the extra-atomic relaxation energy amounts to about 2 eV for all CO levels, with the difference between individual levels amounting to tenths of an electron volt.

In its main features the problem of estimating the value ΔE due to adsorption is analogous to that of estimating the value ΔE for the gas → solid transition for metals. A large number of papers have been devoted to this problem (see references in [75–79]). For this transition the value ΔE is also determined in the main by the magnitude of ΔE_{rel}. Since the value of $|\Delta E_{rel}|$ increases in the series atom → diatomic molecule → solid and $\Delta E_{rel} < 0$, the binding energy of the inner electrons must diminish during these transitions [80], in accordance with experimental results. Note should be made of some patterns in E_{rel} values. The value of extra-atomic or extramolecular relaxation ranges from -1 to -10 eV [77]. The relaxation value becomes large for metals, while for molecules it is usually close to 1–4 eV. For adsorbed systems it increases with the growth of the substrate–adsorbate interaction (Fig. 4.9). In principle, during the gas → solid transition the value of ΔE_{rel} should be connected with some macroscopic values characterizing the solid's polarization capacity. In [81] the value ΔE_{rel} for the gas → solid transition was calculated on the basis of the dielectric's polarization around a positive hole. The calculated values correctly convey the value ΔE_{rel} and the trend of its change for noble gases and some very simple molecules without a dipole moment.

The upper evaluation of the extra-atomic relaxation energy for the free atom → metal transition can be obtained on the basis of the assumption that the screening charge localizes on the atom's upper free orbital. In this case ΔE_{rel} equals half the Coulomb integral of the inner electron's interaction with the electron in the first vacant orbital. Indeed, it turns that values of ΔE_{rel} estimated thus are roughly 50 per cent greater than the value of ΔE [79].

It should be noted that the value ΔE for adsorbed molecules can be assessed also on the basis of thermodynamic cycles without a direct study of the values ΔE_{rel} and E_{ads}. This approach has proved fruitful also when calculating ΔE for the free atom → metal transition [76]. The following cycle was studied in [76]: a metal with the ordinal number Z (Z-metal) is transformed into an atom (Z-atom), in which an inner electron (Z^+) is removed. This latter state, on the basis of the equivalent cores method (see Chapter 1, Section 1.4), is referred to an atom $(Z+1)^+$ with a removed valence electron that is transferred into a neutral atom

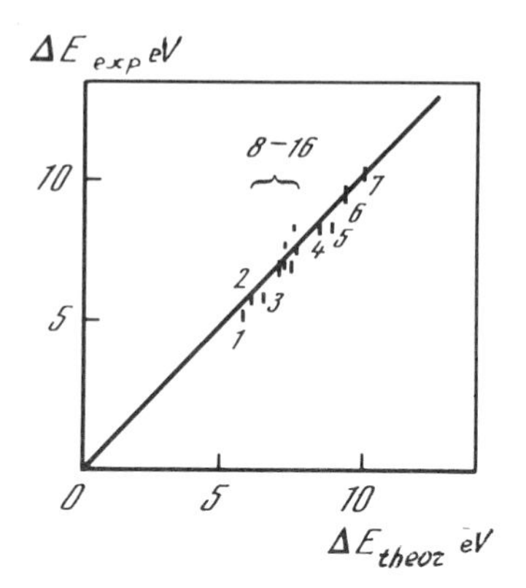

Figure 4.10. Comparison of experimental and calculated ΔE values [see Equation (4.4)]. 1 = Cs; 2 = Rb; 3 = K; 4 = Sr; 5 = Ba; 6 = Ca; 7 = Li; 8–16 = Cd, Hg, Zn, Cu, Yb, Pb, Te, Na, Mg, respectively.

(Z + 1). The atoms (Z + 1) condense in (Z + 1)-metal, after which the (Z + 1)-metal dissolves in Z-metal. On the basis of the equivalent cores method the admixture of the atom of (Z + 1)-metal is regarded as an atom in a metal in which an inner electron has been transferred to the Fermi level. The cycle can be presented as follows:

$$\text{Z-metal} \xrightarrow{E^{Z}_{\text{coh}}} \text{Z-atom} \xrightarrow{E^{A}_{\text{b}}}$$

$$\text{Z}^{+} = (\text{Z}+1)^{+} \xrightarrow{-I} (\text{Z}+1)\text{-atom} \xrightarrow{-E^{Z+1}_{\text{coh}}}$$

$$(\text{Z}+1)\text{-metal} \xrightarrow{E_{\text{imp}}} \text{Z-metal} + (\text{Z}+1)\text{-admixture} \xrightarrow{E^{F}_{\text{b}}} \text{Z-metal.}$$

So for ΔE we have

$$\Delta E = E^{A}_{\text{b}} - E^{F}_{\text{b}} = I + E^{Z+1}_{\text{coh}} - E^{Z}_{\text{coh}} - E_{\text{imp}}. \tag{4.4}$$

The high precision of the model is born out by a comparison of the experimental and calculated ΔE values (Fig. 4.10). An analogous approach was expounded in reference [204] to calculate atomic level shifts in binary compounds.

The problem of studying binding energies of implanted atoms in various matrices also relates to the above material. A detailed analysis of various functions influencing the binding energy of noble gases in metals is presented in [82]. Experimental data on nickel in carbon and Au, Ag, and Cu in SiO_2 is cited in [83, 84].

4.1.4. *Satellites in spectra of adsorbed molecules*

The problem of relaxation during photoionization is closely connected with the structure of satellites in the spectra of adsorbed molecules. The structure of satellites depends on the mechanism of relaxation, that is, on the nature of the screening charge after photoionization. Two main mechanisms are studied in literature [85–90]. According to the atomic or molecular mechanism, the change in the distribution of the charge is connected both with change in distribution in the occupied substrate–adsorbate bonding orbitals (see Table 4.5) and with the transition of a part of the electron density from the substrate's occupied orbitals to the vacant orbitals of the adatom or molecule. In compliance with the plasmon mechanism the screening charge localizes mainly in the substrate, resulting in an excitation of surface plasmons.

Let us study the structure of an inner-level spectrum of an adatom or molecule for two border-line cases when the atomic–molecular or plasmon nature prevails [85]. If the non-diagonal matrix element between the vacant valence level of adatom ε_a and the occupied valence level of metal ε_M^* is designated as V, two energy states are possible for the photoelectron's final state.

$$\varepsilon_\pm = \tfrac{1}{2}(\varepsilon_a + \varepsilon_M) \pm \{[\tfrac{1}{2}(\varepsilon_a - \varepsilon_M)]^2 + V^2\}^{1/2}. \quad (4.5)$$

If the screening electron occupies only the ε_- bonding state, the photoelectron has an adiabatic value of kinetic energy and there is no satellite in the spectrum. But in the general case the screening electron can occupy both $\varepsilon_\pm$ states with a certain probability, and two lines with an intensity proportional to $\beta^2(\varepsilon_\pm)$ are observed in the spectrum:

$$\langle\psi_{\text{ini}}|\psi_\pm\rangle^2 = \beta^2(\varepsilon_\pm) = [1 + (\varepsilon_\pm/V)^2]^{-1}$$

It is important to stress that at $|\varepsilon_a| > |\varepsilon_M|$ the function of the final state ψ is localized on the atom and ψ^+ is localized on the metal, while at $|\varepsilon_a| < |\varepsilon_M|$ there a reverse localization is observed where the function ψ_{ini} is always localized on metal (Fig. 4.11) [85]. As a result, the more intensive line can have both a greater

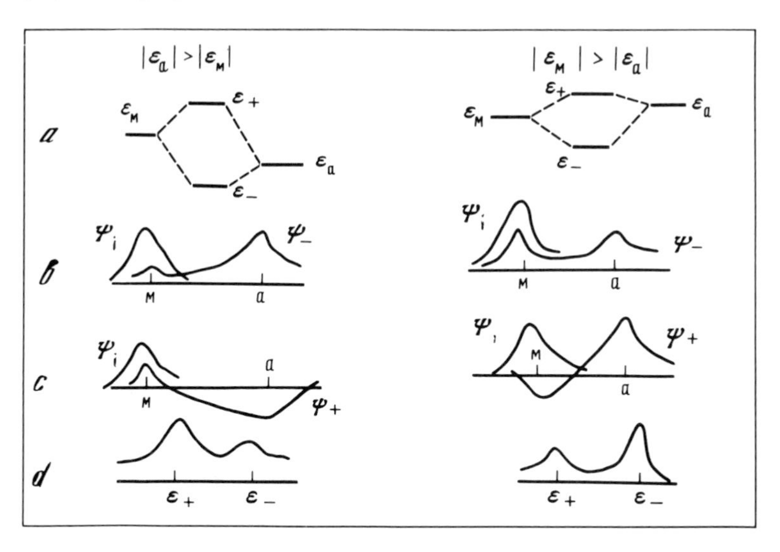

Figure 4.11. Concerning the mechanism of satellite origin. (a) Diagram of levels; (b, c) form of wave functions; (d) type of X-ray photoelectron spectra.

*For simplicity we shall limit ourselves to one level. In reality account should be taken of the metal's conduction band.

and a smaller energy, i.e. the notion of satellite is conditional (Fig. 4.11). Satellites in the spectra of adsorbed molecules (see reference in [89]) are usually explained by this mechanism.

In the case of the plasmon mechanism the energies ε_+ and ε_- corresponding to the photoelectron's final state are equal to

$$\varepsilon_- = \varepsilon_M - \lambda^2/\omega_s, \quad \varepsilon_+ = \varepsilon_M + \omega_s + \lambda^2/\omega_s, \tag{4.6}$$

where λ is the bond parameter of a plasmon with a vacancy and ω_s is the energy of the surface plasmon. Line intensities at ε_- and ε_+ relate as $(1 - a)/a$ where $a = \lambda^2/\omega_s$. So in this case one may expect the presence of plasmon satellites from the substrate in the adatom's spectrum. This was observed for O 1s lines of oxygen atoms adsorbed on aluminium [87].

In the general case one may expect the manifestation of both types of satellite, depending on the parameters λ and V. Ratios and graphs making it possible to evaluate satellite energies and intensities depending on parameters λ and V are presented in [85].

An interesting model of the appearance of strong satellites in the spectra of weakly chemisorbed CO molecules on Cu and Au metals and also of N_2 molecules on Ni is offered in [91]. It is noted that strong satellites appear only in the event of an anomalous dependence of the vibration frequency ν(CO) and $\nu(N_2)$ on substrate's coverage rate (decrease of value ν with the increase of the coverage): for instance, satellites are absent in the systems Ru(001)/CO, Pd(100)/CO, Pt(111)/CO while the frequency ν(CO) increases with the growth the surface coverage. The suggested model explains both the presence of satellites and the anomalous change of the values. In weakly chemisorbed CO and N_2 molecules the 2π orbital of these molecules does not take a substantial part in chemical bonding in the ground state. In the ionized final state the 2π orbital turns out to be below the Fermi level and occupied (atomic–molecular mechanism of satellites, see above). The substrate–admolecule bond decreases with the increase of surface coverage rate. In strongly chemisorbed CO and N_2 molecules the 2π-antibonding orbital has a noticeable population and a decrease of this population increases frequency ν. In weakly chemisorbed molecules the metal–molecule bond is effected at the expense of the donor orbital 5σ, which is a weakly antibonding one for CO and N_2 molecules. Likewise, a lessening of the donor nature of 5σ and growth of the coverage rate of the metal's surface should decrease the value ν(CO) and $N(N_2)$. Reviews of various types of satellite in the photoemission of adsorbed molecules are presented in [71, 92].

4.2. Application of the angular dependence of intensities in studying adsorbed molecules

Electron spectra for molecules adsorbed into surface xy can be obtained for different angles α, θ, and Φ (Fig. 4.12). The intensity I of photoelectrons depends on the matrix element

$$I \sim \langle i|\mathbf{A}\cdot\mathbf{p}|f\rangle^2, \tag{4.7}$$

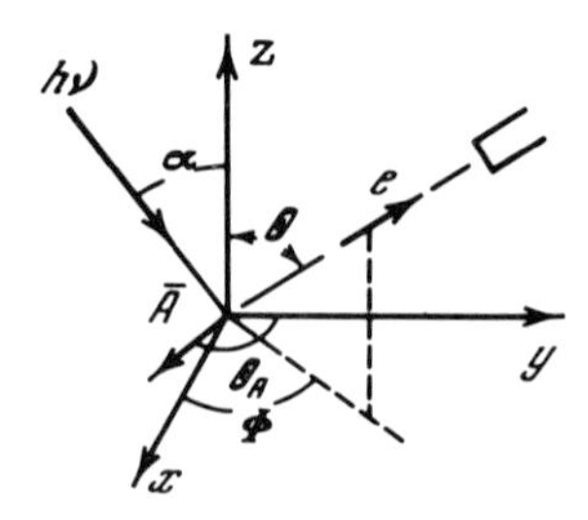

Figure 4.12. Diagram of experiment to study angular dependence.

where i and f are the wave functions of the initial and final states, **A** is the vector potential of the falling radiation, and **p** is the transition operator in 'velocity' form. Since **A** and **p** are vectors, intensity I depends on angles α, θ and Φ and on the polarization of ionizing radiation.

At present the angular dependence I is used extensively to determine the symmetry of the wave function and the geometric position of the molecule on the substrate (see reviews [93, 94]).

4.2.1. *Dependence of* I *on angle* α

This relationship [95–99] is easy to obtain from Equation (4.7) written in the form of a 'length' operator.

$$I \sim |\mathbf{A} < i|e\mathbf{r}|f|^2 \quad (4.8)$$

The dependence of I on α is especially manifest when $\theta = 0$. For I to be other than 0, function f must not have a knot at the point where the detector is positioned, and if axis z is the symmetry axis, f must hold for all the symmetry operations, i.e. must correspond to the full-symmetry representation a_1. Under this requirement, each **r** and **A** must have appropriate symmetry properties because the transition matrix element also must be invariant with regard to symmetry operations and correspond to the full-symmetry representation a_1. Hence, function i must belong to the same irreducible representation as components p_i or r_i. For instance, in the case of point symmetry groups C_{4v} and C_{3v}, states a_1 and e have corresponding to them the polarization of photon z and x, y; in point group C_{2v}, states a_1, b_1, and b_2 have corresponding polarizations z, x, and y. These general statements can be illustrated by two examples: O spectra on metallic Ni [96, 97] and C_6H_6 spectra on Pd [98, 99].

The spectra of the valence levels of the Ni/O system (depending on angle α values marked on curves) are represented in Fig. (4.13). When an O atom is adsorbed on an Ni atom, surrounded with four more Ni atoms, a local symmetry system C_{4v} is formed. We should expect the formation of two valence levels: a_1(O2p$_z$, σ-bond O–Ni) and e(O2p$_{xy}$, π-bond O–Ni). Since non-polarized radiation He(II) is used to excite the spectrum, photon plarization with $\alpha \to 0$ corresponds to x, y and photoemission intensity for a_1 must equal zero. With $\alpha \to 90°$, photon polarization corresponds to z, x, and the intensity of level e must be below that for $\alpha \to 0°$. Levels a_1 and e can be identified on the basis of these

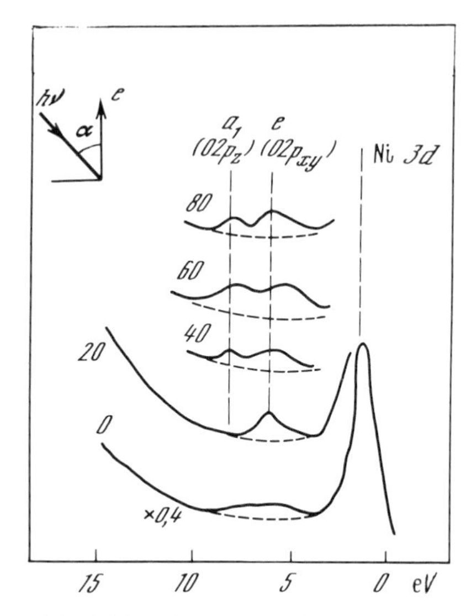

Figure 4.13. He(II) spectra of the $Ni(001)O_2$ system at $\theta = 0$.

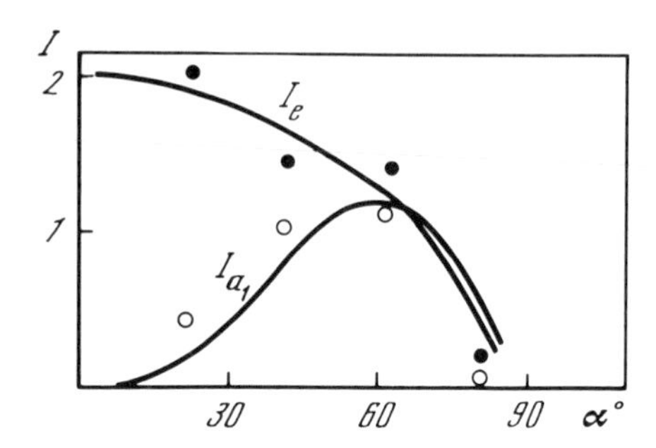

Figure 4.14. Theoretical (curves) and experimental (dots for the e level and circles for the a_1 level) dependencies of I_a and I_e on angle α.

considerations (Fig. 4.13). When experimental data are interpreted, it is also necessary to take account of the reflection of radiation from the substrate, which also depends on α. If this phenomenon is accounted for, we obtain the following equation for intensity from levels a_1 and e:

$$I_{a_1}(\alpha) = |1 + r_p(\alpha)|^2 \sin^2 \alpha,$$
$$I_e \sim |1 - r_p^2(\alpha)| \cos^2 \alpha + |1 + r_s(\alpha)|^2 \tag{4.9}$$

Figure (4.14) shows theoretical and experimental dependences for I_{a_1} and I_e. For small values of α, $I_{a_1} \sim \sin^2 \alpha$ and $I_e \sim 1 + \cos^2 \alpha$. The similar dependence of I on α for small values of α should also be expected in other systems.

Figure (4.15) shows spectra Pd/C_6H_6 with $\alpha = 0 \div 80°$. Functions f of the final states is correspond to the representations a_{1g} and a_{2u} (within the framework of the D_{6h} symmetry for free C_6H_6), which evolve into a_1 (within the framework

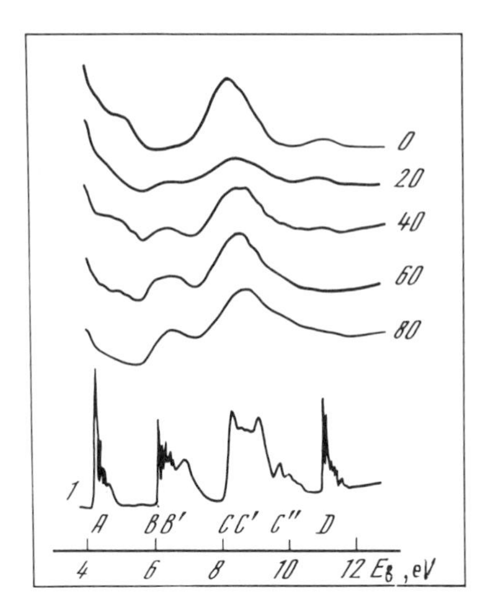

Figure 4.15. Dependence of He(I) spectra of the Pd(100) + C_6H_6 system on angle α. 1 = spectrum of the free molecule C_6H_6.

of symmetry C_{6v} for adsorbed C_6H_6). The results of group theory analysis which determines permitted transitions on the basis of (4.8) are shown in Table 4.6.

This, with $\theta = 0$ and x, y polarization, we can observe levels e_{1g} and e_{1u}, and with z-polarization levels a_{2u} and a_{1g}. Orbitals e_{2g}, b_{2u} and b_{1u} cannot be observed with these polarizations. Non-polarized radiation He(I) was used in [98], so angle $\alpha = 20°$ is close to polarization x, y, and angle $\alpha = 60°$ to polarization z. Since at $\alpha = 20°$ the A and C peaks are strong but D is weak, and since at $\alpha = 60°$ A and C decrease and D grows, with the B peak not in evidence in both cases, the A, C, D, and B peaks correspond to the e_{1g}, e_{1u}, a_{1g} and e_{2g} levels, respectively. The π-level of a_{2u} in the spectrum of the C_6H_6-free molecule is close to B(B), while in the adsorbed molecule this level sinks deeper, into the C zone. Hence, both π-levels of e_{1g} and a_{2u} interact with the substrate in the course of adsorption and increase their binding energies by roughly 0.7–0.8 eV.

Reference [98] studies the dependence of I on angle θ and offers a simple model explaining this dependence. In this model function i is expanded into spherical harmonics (limited to small values of l). Values l and m for final state f and therefore the dependence of I on θ are determined on the basis of the selection rules $\Delta l = \pm 1$ and $\Delta m = 0, \pm 1$.

Table 4.6
Permitted transitions and photon polarization

i							
e_{1g}	e_{2g}	a_{2u}	e_{1u}	b_{2u}	b_{1u}	a_{1g}	f
—	—	z	x, y	—	—	—	a_{1g}
x, y	—	—	—	—	—	z	a_{2u}

The dependence of I on α also was used to interpret the levels of the adsorbed molecules CO, C_2H_2, and C_2H_4 [100, 101].

4.2.2. *The dependence on θ and α*

This dependence was used in [102–104] to determine a coordination type of the CO molecule onto Ni and Pt. The intensity ratio $4\sigma/(1\pi + 5\pi)$ was measured depending on angle θ, while the angle between the direction of the photo fall and the photoelectron exit remained unchanged. This dependence for different types of orientation—M–*CO*, *M–OC and CO*–parallel to metal surface M—was determined on the basis of computation [105]. It turned out that experimental data accord best of all with the M–CO coordination.

The dependence of I on θ and α is also used to determine the order of the 1π and 5σ levels in the adsorbed CO molecule [102, 106]. To this end, reference [102] employed the analogy in dependence 5σ and 4σ, or the quantitative results of computation [105]. In fact, the dependence on α examined earlier could also be used. Figure (4.16) shows spectra of the Pt/CO system for different angles θ, with the angle between directions h and e being constant ($62°7'$) [102]. The intensity of the π-level is lowest at $\theta = 0$ and highest at $\theta \simeq 60°$, if we proceed from the M–CO coordination and remember that $\theta \approx 60°$ corresponds to $\alpha \approx 0$, i.e. x, y polarization. It is noteworthy that the 5σ level in the free CO molecule is higher than the 1π level by 2.9 eV, 0.5 eV and higher in $Ni(CO)_4$, but 0.5–1.5 eV lower than the 1π level in the adsorbed CO molecule [58, 100, 102, 106, 107]. This indicates that the 5σ level plays the role of electron pair donor in these systems.

4.2.3. *The dependence of* I *on the final state* f

This can also be used to determine the mode of coordination, the character of the electron level, and the substrate composition. Resonances in the cross-sections of photoionization of the adsorbed molecule on the substrate can be used for the purpose. Let us first consider the utilization of the resonance structure [107, 108].

Usually as hν grows, the cross-section of photoionization diminishes for all

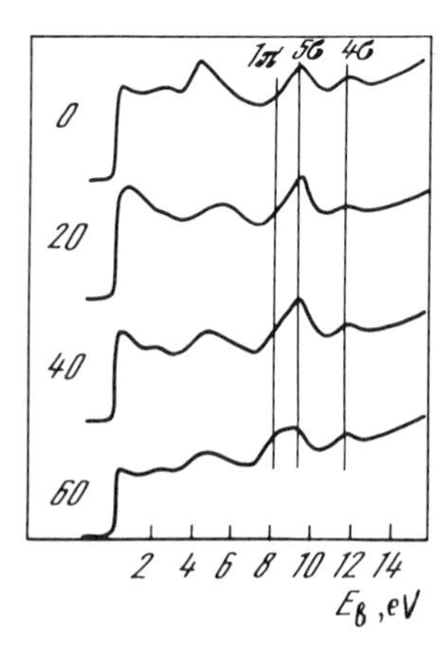

Figure 4.16. Dependence of the He(II) spectra of the Pt(111)/CO system on angle θ (figures next to the curves).

levels at E_{kin} $1 \div 5$ eV for the photoelectron. However, with small E_{kin} values (up to 20 eV), the cross-section of the photoionization level sometimes grows, and since this phenomenon occurs within the narrow limits of E_{kin}, this maximum is called the resonance maximum. This resonance can be viewed as a quasi-stationary state in the continuous spectrum or as a result of the resonance amplification of the amplitude of the photoelectron wave function due to scattering on atoms of the molecule.

Reference [97] used the following computation result [105] to analyse the dependence of the resonance intensity of CO molecules adsorbed by Ni upon the angles hν and photon polarization. The final resonance σ-state with very intensive photoemission is possible for initial σ-states only with z polarization and for initial states only with x, y polarization. The results [107] for the dependence of I on hν for some angle values are supplied in Fig. (4.17). It shows that resonances occur at $\theta = 0$, when the **A** vector has a component at right angles to the xy surface (see Figs 4.12 and 4.17b). There are no resonances with small θ values if **A** lies in the xy plane (see Fig. 4.17a), nor at the angles shown in

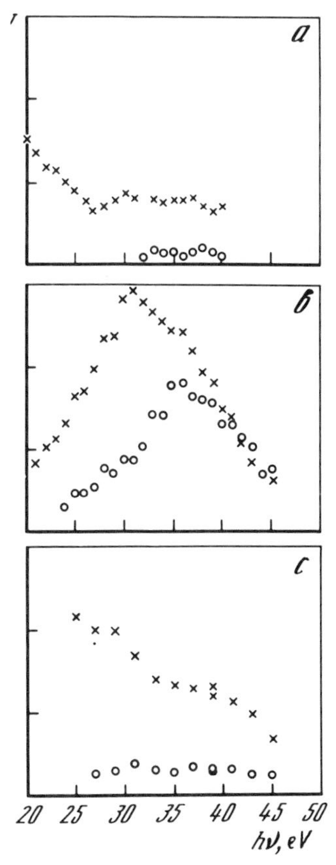

Figure 4.17. Dependence of intensities of the 4σ-levels (dots) and $1\pi + 5\sigma$ (crosses) for different angles on the energy hν. $a - \alpha = 0°$, $\theta = 8°$, $\Phi = 90°$, $b - \alpha = 45°$, $\theta = 0° = \Phi$; $\theta - \alpha = 45°$, $\theta = 63°$, $\Phi = 90°$.

Fig. (4.17c). This prompts the conclusion that the CO molecule is adsorbed perpendicular to the surface and parallel to the z axis. The quantitative analysis of the dependence of intensity of the 4σ-level on angle θ at $h\nu = 35$ eV (which corresponds to resonance) demonstrated that the deviation of the CO axis from perpendicular to the surface is not more than 5°. To determine the coordination method (Ni–CO or Ni–OC), reference [107] also used the dependence of the 4σ-level resonance intensity on different coordination types for angles $\alpha = 45°$ and $\theta = 0°$ and established that CO is coordinated through the C atom. The character of the photoionization cross-sections of the oriented CO molecule depending on $h\nu$ was studied in detail in [109]. It was demonstrated that with sufficiently large $h\nu$ values (> 100 eV), oscillations of the photoionization cross-section ought to occur, which depend on inter-atom distances in CO. This phenomenon makes it possible to determine the characteristics of the adsorbed molecule.

Adsorption of CO perpendicular to the surface of metal is not a general rule: it was established in reference [60] on the basis of a study of angular dependence that in the case of CO adsorption of Pt(110) the axis of CO deviated from the perpendicular to the surface by 26°. A deviation of CO from normal was confirmed for this crystal plane in [110]. It shall be noted that the CO molecule adsorbs on Pt(111) and Pd(111) perpendicular to the surface [111, 112]. A study of the angular dependence of N 1s peaks in the Ni(100)/N_2 system showed that the N_2 molecule, too, was adsorbed perpendicularly to the Ni plane. The value N 1s of N atom connected with Ni turned out to be greater by 1.3 eV than the N 1s value of the N end atom [113].

Let us consider the dependence of inner-level photoelectron intensity in the adsorbed molecule on azimuthal angle Φ. Usually, this dependence is weak or altogether absent [100, 114, 115]. However, it has by now been found for the Se 3d and Te 4d lines in the Ni(001)/Se Te systems and the O 1s line in the O/Cu(001) system [116, 117]. This dependence is observed at large θ angles and is connected to photoelectron scattering on the substrate. The computation and experiment [116] demonstrated that maximum intensity is found in photoemission along the projection of the adsorbed atom–substrate atoms bond. This makes it possible to determine the point symmetry group of the adsorbed atom–substrate system. In principle, the same phenomenon can be studied by varying $h\nu$ and thus E_{kin} of the photoelectron, which yields the dependence of the defraction pattern both on Φ and on $h\nu$ (see theoretical discussion in [118, 119, 120]).

It is likewise possible to study the scattering of photoelectrons resulting from the inner levels on the atoms of the molecule itself. Reference [116] analyses the diffraction of C 1s photoelectrons on the atoms of the Ni(001)/CO system and demonstrates that the axis of the CO molecule is perpendicular to the surface. In general, the diffraction of inner-level photoelectrons of the adsorbed molecules can be used to determine inter-atom distances. The influence of molecular oscillations on the diffraction of photoelectrons was studied in [121].

We have considered above variations in the spectra of the adsorbed molecule, but scattering is also substantial for substrate photoelectrons. It has been demonstrated, in particular, that the incoherent scattering of substrate photoelectrons on adsorbed molecules results in the weakening of the Cu(111)/Cs system [122].

Both the theory and technics of experiments to study photoelectron diffraction and its dependence on kinetic energy and angles are now rapidly progressing. The first papers containing a direct definition of surface structural characteristics have been published [123]. It appears that along with EXAFS this method will become the main one in studying surface structure.

We have considered above adsorbed molecules. The case in which reaction occurs in the top most layer of a solid with the formation of a monolayer of atoms or chemical groups bound with conventional strong chemical bonds to the substrate is not basically different. This occurs, in particular with the implanted complexes on a solid surface. Let us take as an example the Si(111) + Cl systems. The non-saturated bonds of the top monolayer of Si atoms in such systems are saturated through the formation of Si–Cl bonds. Photoelectron spectra of such systems are considered in [124–127].

After Si(111) reactions with Cl, there appear in the photoelectron spectrum two additional peaks with a shoulder of about 5 and 8 eV. These peaks correspond to Cl 2p orbitals. In principle, two options for the bonding of Cl to the surface are possible: a covalent bond (Cl atom over Si atom) and a ionic bond (Cl atom in the centre of three Si atoms). A detailed spectra analysis demonstrated that the first possibility holds for Si(111) (2×1) + Cl. The analysis is as follows. In accordance with the above considerations, the a_1 symmetry levels corresponding to the Si–Cl bond, ought to yield a far smaller intensity for s-polarized ionizing radiation (with the polarization vector **A** perpendicular to the photoelectron exit path) than for p-polarized radiation (**A** lying in the photoelectron exit plane). Since this dependence was actually found for the peak with an energy of about 8 eV, it should be linked to the Si–Cl bond, formed by the Cl $3p_z$-orbital. The peak with an energy of about 5 eV belongs to the Cl $3p_{x,y}$ states.

4.3. Study of catalysts

X-ray photoelectron methods make it possible to determine the chemical composition of the surfactant phase of a catalyst, the reaction of the catalyst with the carrier, the dependence of the catalyst composition on temperature, the course of the reaction, and the action of various inhibitors. It is also possible to study the processes of surfactant layer formation, catalysts contamination, etc. (see reviews [3, 128] and bibliographsies).

We will consider in below only questions related to the quantitative X-ray photoelectron analysis of the catalyst surface.

4.3.1. *Oxides and oxides on carriers*

The importance of determining the composition of the catalyst surface is obvious. The possibilities of X-ray photoelectron spectroscopy in this field are illustrated by Fig. (4.18) [129], which represents changes in the relative intensity of the Bi $4f_{7/2}$/O 1s and Mo $3d_{5/2}$/O 1s lines in the catalysts Bi_2MoO_6, $Bi_2Mo_2O_9$, and

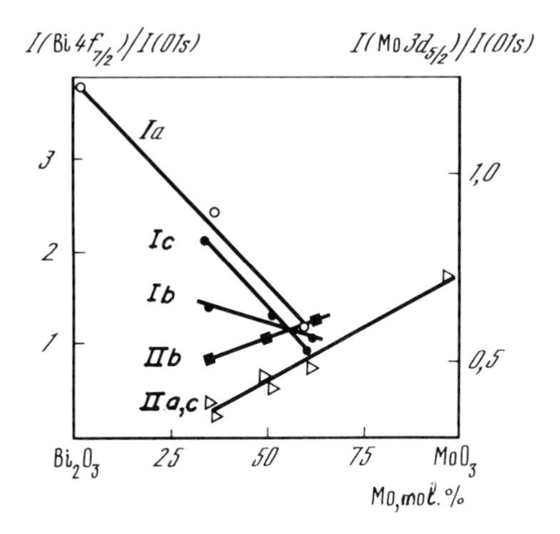

Figure 4.18. Changes in the intensities of the Bi 4f(I) and Mo 3d(II) X-ray photoelectron lines versus the O 1s line after different treatments of the catalyst.

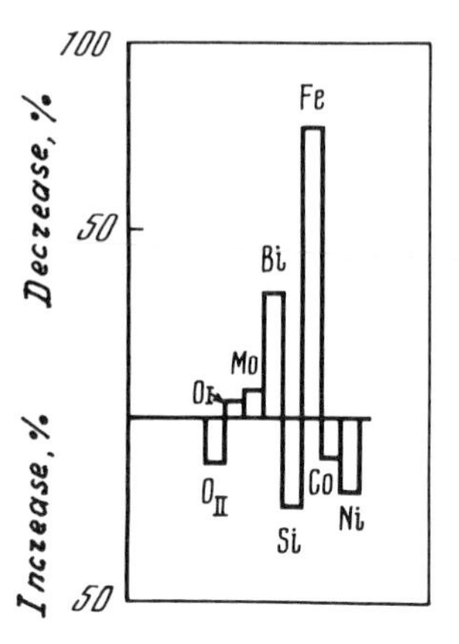

Figure 4.19. Changes in the surface composition of the catalyst after reaction. O_I = oxygen atoms in metal–oxygen bonds; O_{II} = oxygen in Si–O bonds.

$Bi_2(MoO_5)_3$. Figure (4.18) shows the lines in the initial catalysts (Ia, IIa), after heating the catalysts for 10 h in vacuum at a temperature of 470° C (Ib, IIb), and after treatment with the reactive mixture $C_3H_6 + O_2 + N_2$ (Ic, IIc). It is clear from the figure that heating in vacuum increases the concentration of molybdenum on the surface (due to structural changes and also through diffusion of Mo to the surface and evaporation of Bi from the surface). Under conditions of propylene catalytic oxidation, the composition of the catalyst surface phase coincides with the volume.

Figure (4.19) [130] shows the change in the catalyst 50% $Ni_3CO_5Fe_3ViPK_{0.1}Mo_{12}O_{52.5}$/50% SiO_2 after the catalytic oxidation of propylene. Significantly, changes in the composition of the surface in this case are caused by the reaction rather than by a rise in temperature. On the other hand, in the case of $Cr_2O_3CuOCaO/Al_2O_3$ catalysts [131], a drastic change in the Ca surface content is related to temperature: as the temperature rises from 600 to 1000° C, the Ca/Al ratio grows 2.5 times.

A particularly large surface enrichment ratio compared to the bulk was found in the case of Cr_2O_3 in CoO: surface concentration grows as Cr concentration decreases in bulk and with a 0.048 atom concentration of chrome in bulk amounts to 100 [132]. For further examples of the dependence of surface composition of catalysts on the duration of reaction or various treatment, see [133–138].

The influence of admixtures on the composition of the catalyst surface has been studied in [139, 140]. As the concentration of NaH_2PO_4 changes from 0 to 5% in the V_2O_5–MoO_3 catalyst, the ratio of the surface concentrations V/Mo changes fourfold, while the same ratio in bulk remains unchanged. Added to the V_2O_5–MoO_3 solid solution, different oxides also change the V/Mo ratio on the surface. (Other noteworthy examples of the quantitative analysis of catalyst surfaces are supplied in [141–149]).

Let us consider some special aspects of the analysis of catalyst surface composition. Difficulties involved in the quantitative analysis here are explained by the fact that the methods of analysis (see Chapter 2) have been developed only for the homogeneous distribution of elements on the surface, while in the case of catalysts we have to reckon with the influence of porosity, particle size, and concentration gradient.

If a surface is porous, the substance being studied may occur on the surface of pores and contribute to catalysis, but it will not be found by XPS because this method only records the composition of the outer surface layer. Reference [150] established, for instance, that the Mo/Si and Co/Si intensity ratios in the Mo/SiO_2 and $CoMo/SiO_2$ catalysts are virtually independent of Mo and Co content at Mo and Co concentrations below 12%. However as Mo and Co concentrations increase above this value, Mo/Si and Co/Si intensity ratios rise dramatically. This is explained by the parallel growth of Co and Mo particle size with an increase in the Co and Mo content to 12 per cent. When the Co and Mo content of Co and Mo oxides exceeds 12%, the metals are found mainly on the surface of carrier particles rather than in pores.

An interesting phenomenon is discussed in [151], which analyses the (5–15% CrO_3)/SiO_2 catalyst. As the catalyst is heated, the reduction $CrO_3 \rightarrow Cr_2O_3$ takes place; moreover, at a temperature of 500° C, within a narrow range of about 20° C, there is a leap in the intensity of the Cr 2p line, and an O 1s line with an energy of 530.7 eV (bound to Cr_2O_3) appears. The authors explain this phenomenon by the formation of a finely dispersed α-Cr_2O_3 phase. When the substance is heated more, to 550° C, the above effect is not observed: it can be presumed that agglomeration of Cr_2O_3 particles occurs at this temperature. Neither does the effect occur when a sample with finely dispersed α-Cr_2O_3 is heated to 300° C for 4 h. Apparently, this phenomenon is hard to reproduce. Reference [152], which studied the same catalyst, notes a constant decrease in the intensity of the Cr 2p line during heating due to the agglomeration of Cr_2O_3, but does not mention any effect at around 500° C.

The influence of particle size on observed intensity will be discussed in greater detail below. The next problem to be discussed is the surface to bulk concentrations gradient (see Chapter 2 for general theoretical discussion). Reference [153] considers various evaluations of nickel concentrations in the surface layer of the NiO/SiO_2 catalyst, taking into account the concentration gradient.

$$X_t = X_v + (X_0 - X_t)\exp(-t/n),$$

where t is the distance from the surface, and X_0, X_v, and X_t are nickel concentrations on the surface, in the bulk, and at a distance t from the surface. The value n characterizes the gradient of concentrations. Estimates show that the enrichment of the topmost surface layer of the catalyst with nickel oxide depends on X_v and that with small nickel concentrations, the X_0 composition of the topmost layer may markedly differ from the average composition of the surface layer as determined by XPS.

Reference [154] analyses the surface composition of MgO–ZnO solid solutions using the Zn 2p, Zn 3s and Zn (MM) lines. These lines are characterized by photoelectron exit depths of $\lambda = 0.815$ and 1.3 nm. It also was established the surface layer was enriched with zinc oxide and that the degree of enrichment, as was to be expected, grows as λ diminishes. The difference $\Delta = \bar{X}_\lambda - X_v$ was determined for every line, where $\bar{X}_\lambda$ is the average zinc content in the analysed surface layer and X_v is the zinc content in the bulk. According to theory, enrichment takes place in one or two surface monolayers simultaneously. The intensity contribution of this layer or layers to the total intensity is easy to compute.

Bearing in mind that

$$yX_s + (1 - y)X_v = \bar{X}_\lambda \tag{4.10}$$

we will have for the X_s concentration in the topmost layer

$$X_s = X_v + \Delta/y \tag{4.11}$$

Calculation of X_s according to this equation produced close results for different lines. Moreover, close results were also obtained in the computation of X_s^{Zn} and X_s^{Mg} according to the equation

$$X_s^1/X_s^2 = (X_v^1/X_v^2)\exp(\sigma_2 - \sigma_1)S/RT, \tag{4.12}$$

where σ is the surface tension of solution components, and S the specific mole area. The above equation takes account of changes in surface concentration depending on σ_i and X_v. To date a large number of experimental and theoretical papers on surface segregation in alloys and solid solutions have been published (see reviews [155–159]).

It should be noted that the dependence of the intensities of X-ray photoelectron lines on composition for CuO–MgO, CoO–MgO, and other catalysts is discussed in [160]. It was established that I(Cu $2p_{3/2}$/Mg ls) has a linear relationship to the Cu/Mg ratio in bulk in the CuO–MgO system in solid solutions and that the proportionality coefficient alters with a transition to the multi-phase system.

4.3.2. *Zeolite catalysts*

Quantitative proportions on the surface of zeolite catalysts are discussed in a number of papers ([3, 161–173]: see also review [174]).

Table 4.7 [162] supplies data for the relative concentrations of elements in the NaX and RhX catalysts. It shows that the surface Al/Si and O/Si ratios are just one half of such ratios in the bulk medium. Sodium concentration on the surface is low. Figures on Al/Si concentrations are similar to those cited in references [161, 163, 164, 170] but disagree with those in [166–168]. Review [3] explains this controversy by the inadequacy of analysed specimens. It therefore seems that the impoverishment of the surface of zeolite catalysts by Al atoms is still an open question.

Table 4.7 shows that the RhX surface is enriched with Rh. An even higher degree of enrichment of the surface with the transition elements Cu, Co, Fe is recorded in [161, 163]. The surface of the RuNaY zeolite is poor in Ru (if there is no oxygen in the system) after reduction with hydrogen or after the reaction of CO and H_2 conversion into CH_4. However, the presence of as little as 0.02% oxygen is enough to isolate dispersed RuO_2 on the surface [170].

X-ray photoelectron spectra have proved useful in the study of changes in the composition of zeolite surfaces under various treatments. Study [175] established for the Ni–Cu zeolites that treatment with hydrogen increases Ni/Si and Cu/Si intensities, testifying to the diffusion of copper and nickel atoms to the surface, where they form an alloy, according to alteration of magnetic susceptibility. Oxidation treatment decreases Cu/Si intensities, though the absolute intensity of the Si 2s line also decreases by half. The results in [175] are explained, on the one hand, by the suspension of metal atoms on the surface, which ought to reduce the intensity of the Si 2s line through the screening of Si atoms, and on the other hand, by the decrease of the Cu/Si surface ratio due to the diffusion of the Cu^{2+} ion into the zeolite bulk. The reduction of RhX zeolite with hydrogen [162], conversely, decreased the intensity of the Rh line, which is explained by the agglomeration of metallic rhodium.

Reference [167] (Table 4.8) describes a rather complex relationship between Ag surface concentration in AgCaNaAl zeolite and Ca content and the treatment of the surface with olefins. As Ca content grows, Ag concentration on the surface

Table 4.7
Relative proportions of various elements at surface and in bulk medium for Na*X* and Rh*X* [162]

Substance	Na*X*		Rh*X*		
	Surface	Bulk	Surface	Surface[a]	Bulk
Na	0.33	0.80	0.18	0.31	0.80
Al	0.42	0.80	0.46	0.48	0.80
Si	1.0	1.0	1.0	1.0	1.0
O	1.9	3.6	1.8	1.9	3.6
Rh	—	—	0.047	0.050	0.01

[a] After heating at 475 K for 18 h.

Table 4.8
Relative intensities of spectral lines in AgCaMaAl zeolite with 8% Ag

Ca, %	Before treatment		After treatment	
	Ag $3d_{5/2}$/O 1s	Ag $3d_{5/2}$/Al 2s	Ag $3d_{5/2}$/O 1s	Ag $3d_{5/2}$/Al 2s
0	1.1	13.5	0.50	6.0
13	0.42	5.5	0.58	6.0
16	0.50	5.5	0.38	5.3
20	0.18	2.0	0.31	3.9
34	0.21	2.2	0.36	4.7

decreases to a point where it is close to the average bulk concentration (8%), while the treatment of the surface with olefins can both decrease and increase the Ag concentration on the surface. These findings are explained by the influence of pore size, which depends on Ca concentration, upon the reduction of Ag^+ to Ag and upon diffusion processes in zeolite.

4.3.3. *Metal-deposited catalysts*

Here are a few examples to illustrate the application of X-ray photoelectron surface analysis to determine various characteristics of catalytic processes.

Study [176] established that the Na diffusion from the substrate to Ag or the effect of an organic deposit on Ag drastically reduces the activity of that catalyst for ethylene oxidation. Reference [177] analysed the Ni/α-Al_2O_3 catalyst used in natural gas reforming. It established that there was a decrease in the intensity of the Ni lines in the catalyst used, caused by the baking of nickel particles. As would be expected, the intensity of the Ni 2p line decreases more than that of the Ni 3p line because the photoelectron mean free path in the latter case is bigger. The same study unexpectedly discovered an accumulation of Pb on the surface of the operating catalyst, which contaminated the latter. A thorough study of the entire system revealed that the source of Pb was a tap through which distilled water was being fed, and the Pb content in the tap material was a mere 1.5–2.5%. The metal tap was replaced with a Teflon one, and the catalyst no became longer contaminated with lead [128].

Study [178], unlike earlier papers, which discussed catalytic inhibitors, concentrated on promoters of the Ag catalyst used in ethylene oxidation. It was already known that the yield of ethylene oxide is drastically increased by γ-radiation. An X-ray photoelectron analysis showed that γ-radiation increases the Ca content of the surface, thus contributing to the formation of O_2^- groups.

Now let us consider the influence of particle size on the relative intensities of metal I_M and carrier I_c [179–189]. According to theory (see Chapter 2), we have for metal with an average thickness t, covering the kth part of the carrier surface:

$$I_M/I_c \sim I_{1M}k(1-e^{-t\lambda_M})/I[1-k(1-e^{-t/\lambda'_c})] \tag{4.13}$$

If $\lambda'_M \approx \lambda_c$ and $k = m/\rho t = a/t$, where m is the mass and ρ the density of the deposited metal per unit of the carrier area, we obtain

$$I_M/I_c \sim (a/t)(I-e^{-t/\lambda})/[I-(a/t)(I-e^{-t/\lambda})]. \tag{4.14}$$

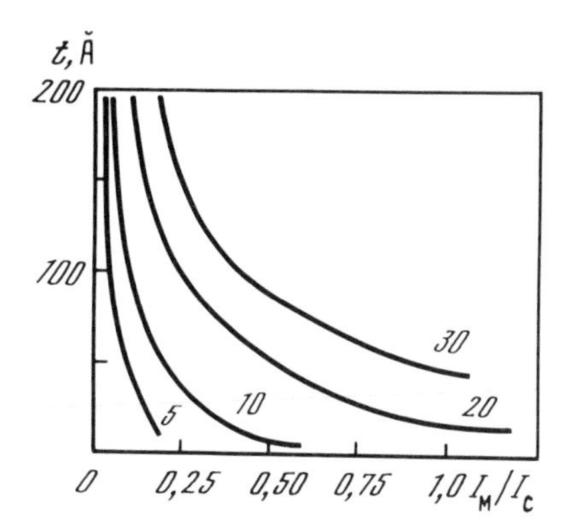

Figure 4.20. Dependence of I_M/I_c on t with different a values (figures next to curves). See Equation (4.14).

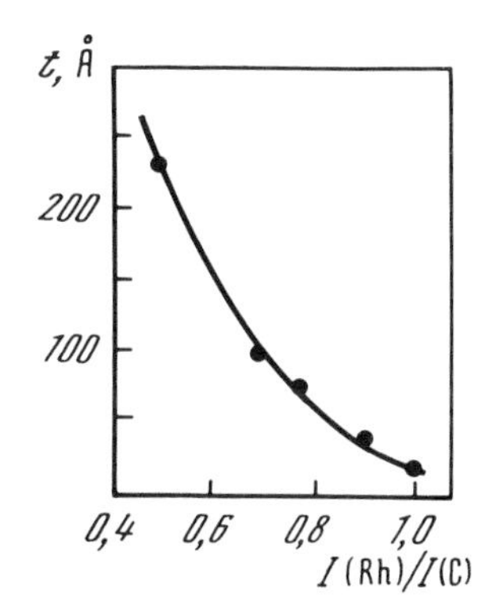

Figure 4.21. Dependence of the Rh/C ratios of intensities of X-ray photoelectron lines on the size t of the Rh crystals. Dots represent experimental data.

Figure (4.20) shows the I_M/I_c function depending on t, with $\lambda = 2.5$ nm and $a = 0.5$, 1.0, 2.0, and 3.0 nm. As expected from general considerations, I_M/I_c diminishes as t groups. At present only one quantitative experimental result of this kind is known [179] (Fig. 4.21). As the size of rhodium crystals grows, the intensity of Rh/C in the 12.5 Rh/C catalyst diminishes. (Independent measurements of t are not usually found in most studies.)

In the general case, caution must be taken in drawing conclusions about the growth of t from a decrease in I_M/I_c intensities because this process, for instance, with a rise in temperature, can also be caused by the diffusion of metal into the bulk of the catalyst or by the formation of a surface film on metal particles. It should be further noted that strictly speaking, the I_M/I_c ratio depends non-linearly on m (or a). However, it is clear from Fig. (4.20) that the I_M/I_c value depends approximately linearly on a (or m), so researchers usually note the linear I_M/I_c dependence (see bibliography in [3]).

Since a linear relationship really exists (with an accuracy of up to k^2) between

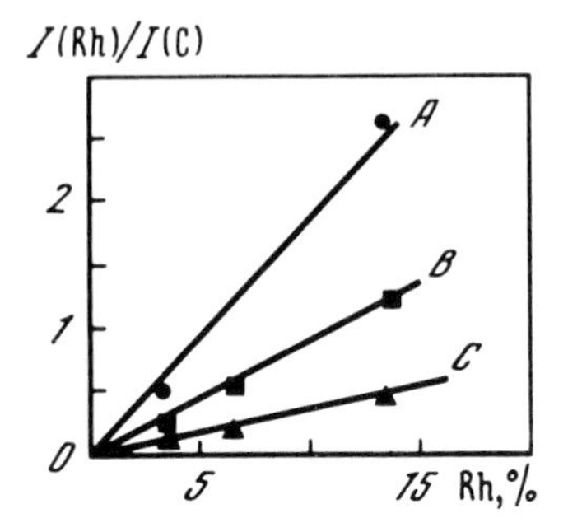

Figure 4.22. Dependence of the Rh/C line intensity ratio on Rh content with different carrier pore sizes.

I_M/I_c and m for small values of k, the extent of deviation from that linear dependence can be used for the qualitative evaluation of k or t.

The above discussion applies only for the case of one external surface exposed to X-ray photoelectron study. If there are pores (inner surface) in the carrier, the I_M/I_c intensity ought to decrease in inverse proportion to the size of the pores in which deposited substances may occur. Figure (4.22) [180] shows data for the Rh/C catalyst; there are virtually no pores in carbon in case A, while the average pore size for B and C is 1.1 and 5.3 nm, respectively.

Generally speaking, the dependence of the I_M/I_c intensity on the size and shape of the carrier can be used to determine the latter. This method is suggested in [185]. The essence of this approach is as follows. If value k is considered small we obtain, on the basis of (4.14):

$$R \sim I_M/I_c \sim AI_{1M}k(1 - e^{-t/\lambda_M})/I_{1c}. \tag{4.15}$$

Factor A is introduced to account for the distribution of the carrier on the inner and outer surfaces. The $1 - e^{-t/\lambda}$ multiplier in (4.15) is connected with the assumption that adsorbed (metal) particles are flat. If they are presumed spherical or semi-spherical, the multiplier will be different. We have in the general case

$$R \sim AI_{1M}kG/I_{1c}, \tag{4.16}$$

where G accounts for the size and shape of the particles.

Factor R is determined with regard to a catalyst with the same value of A and the known factor G_0. Study [185], for instance, used data for a catalyst with a monoatom platinum layer. (This need for standardization of R detracts from the efficiency of the method.) Thus,

$$R = R/R_0 \sim kG/k_0G_0. \tag{4.17}$$

This equation makes it possible to determine the dependence of R_{st} on particle size and shape and, using experimental R_{st} values it is possible to calculate particle size for various assumed shapes. Computation for the Pt/SiO_2 catalyst with an assumed semi-spherical particle shape coincided with the findings of an electronic microscopy analysis.

Deposited metals reveal both positive and negative line shifts as compared to compact metals (see references [190–193]). Usually metal lines shift in the direction of greater binding energies as compared to compact metal. It is often attempted to interpret this as a result of interaction between the metal and the carrier. But it was shown in reference [191] that this shift correlates with the size of the Pt and Rh particles deposited on Al_2O_3 and TiO_2: the smaller the particle the greater the shift. The shift is explained by a change in inter-atomic relaxation in small particles [187]. In the case of small particles, along with the shift, there is also a lessening of line assymetry as a result of the lesser vacancy screening by valence electrons and a growth of line width because of the attenuation of plasmons on the border of particles [192].

It was shown, however, in [193] that the Fermi level of small gold particles on a non-conducting sublayer, just as the Au 4f, line shift towards greater energies. This phenomenon is explained not by relaxation effects, as in [191], but by the presence of a single positive charge on the cluster in the final state of photoemission. The shift ΔE approximately equals $E \sim e^2/r$, making it possible to assess the size of cluster r. This explanation removes the contradiction between the negative shift of the Au 4f line of atoms on the surface and the positive shift of this line in small clusters of gold. Indeed, in small Au clusters the number of surface atoms is relatively large and the line shift in the absence of charging must be negative.

4.3.4. *Alloys*

The specific characteristics of an X-ray photoelectron study in this case lie in the possibility of ion sputtering to obtain information about the distribution of elements in depth (see Chapter 3). The following are a few examples of the analysis of the composition of the catalyst surface on the basis of alloys or compounds closely related to them.

References [194, 195] show that, as the heating time of the Ni–Zr–H catalyst grows, the surface nickel content increases and simultaneously catalytic activity rises (Fig. 4.23).

An analysis of the composition of the surface of different intermetallides NiZr using the angular dependence of intensity and ion sputtering showed (see Chapter 2, Section 2.3) that the active layer of nickel on the surface of the catalyst is formed as a result of the diffusion of nickel atoms through a layer of zirconium oxides. The presence of oxygen, needed for the formation of zirconium oxide, is an essential precondition for obtaining an active Ni layer, since heating alone does not activate the catalyst.

The correlation between surface composition and catalytic activity has been noted in many papers, concerning different catalysts (see e.g., [90, 139, 140, 144, 183, 196–198]). Reference [197], for instance, shows that the activity of the Pd–Rh alloy in the dehydrogenization of cyclohexane is related to Rh content on the surface. However, there can be no such simple correlation between the surface composition and catalytic characteristics (see e.g., [199] for the rhenium copper catalyst).

The surface composition of the alloy may noticeably vary, depending on the composition of the gaseous environment and the reductive or oxidizing

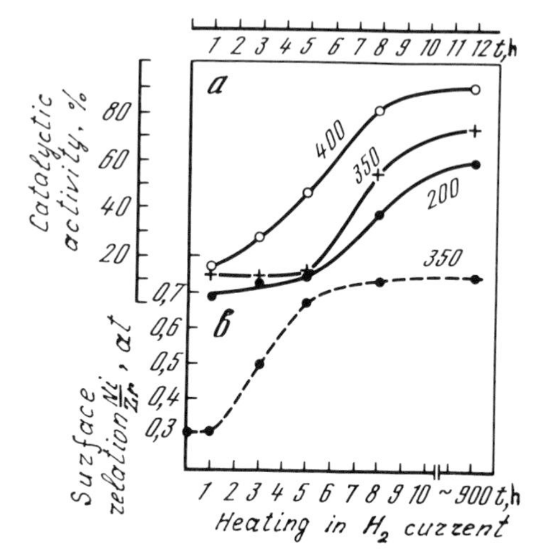

Figure 4.23. (a) Dependence of catalytic activity (degree of transformation of initial product) on catalyst heating time at different temperatures, °C.
(b) Dependence of the Ni/Zr atomic ratio on the surface upon heating time at 350 °C.

treatment of the surface. For instance, the surfaces of Pt–Sn, Ni–Al, Pd–Al, Pd–Cu alloys (see references in [3] and reference [200]), when treated with hydrogen, are enriched with the second component. As a consequence, the surface composition of alloy catalysts can markedly change in the course of reaction. For instance, in [201] it was found with the help of ion sputtering that the surface of the Pt–Rh grid for the catalytic oxidation of NH_3 is covered with an enriched Rh_2O_3 layer. The thickness of this layer is about 2 μm. Reference [202] established that Ti atoms in the hydrogenization of TiFe drift towards the surface, forming clusters of Fe atoms, which catalyse the H_2 adsorption and cause irreversible changes in magnetic properties during hydrogenization.

A detailed discussion of the catalytic characteristics of Pt–Pd, Pd–Ni, Pt–Au and Ni–Cu alloys, depending on their surface composition [203], has established a relationship between surface segregation, catalytic activity, and catalytic mechanisms.

References

1. Brundle, C. R. (1975) Elucidation of surface structure and bonding by photoelectron spectroscopy. *Surf. Sci.* **48**, 99–136.
2. Joyner, R. W. (1977) Electron spectroscopy applied to the study of reactivity of metal surfaces: a review. *Surf. Sci.* **63**, 291–314.
3. Minachev, Kh. M., Antoshin, G. V., and Shpiro, Y. S. (1978) Application of photoelectron spectroscopy in studies of catalysis and adsorption. *Usp. Khim.* **47**, 2097–2133 (in Russian).
4. Gomoyunova, M. V. (1977) Photoelectron spectroscopy of adsorbed atoms and molecules: a review. *Zh. Tekhn. Fiz.* **47**, 673–708 (in Russian).

5. Fuggle, J. C. (1977) XPS in ultra-high vacuum conditions. In: Handbook of X-ray and ultraviolet photoelectron spectroscopy. Heyden, pp. 288–302.
6. Hagstrum, H. D. (1976) The determination of energy level shifts which accompany chemisorption. *Surf. Sci.* **54**, 197–209.
7. Carley, A. F., Joyner, R. W., and Roberts, M. W. (1974) Reference levels in photoelectron spectroscopy. *Chem. Phys. Lett.* **27**, 580–582.
8. Sitte, J. (1976) Relation between reference levels, work functions and contact potential differences in photoelectron spectroscopy. *Chem. Phys. Lett.* **42**, 131–132.
9. Menzel, D. (1978) Photoemission and electronic properties of surface. Wiley, Chichester 381–408.
10. Broughton, J. Q. and Perry, D. L. (1978) Electron binding energies in the study of adsorption by photoelectron spectroscopy: the reference level problem. *Surf. Sci.* **74**, 307–317.
11. Yates, J. T., Erickson, N. E., Worley, S. D., and Madey, T. E. (1974, 1975) The physical basis of heterogeneous catalysis. Battele Institute of Materials Science Colloqium, Gstaad (1974), New York (1975), pp. 75–85.
12. Fukuda, Y., Honda, F., Rabalais, J. W. (1980) XPS, UPS and TDS study of adsorption, oxidation and hydrogeneration of CO on rhenium. *Surf. Sci.* **93**, 338–350.
13. Joyner, R. W. and Roberts, M. W. (1974) Oxygen (1s) binding energies in carbon monoxide adsorption on metals. *Chem. Phys. Lett.* **29**, 447–448.
14. Fuggle, J. C., Umbach, E., Menzel, D. *et al.* (1978) Adsorbate line shape and multiple lines in XPS, comparison of theory and experiment. *Solid State Commun.* **27**, 65–69.
15. Luftman, H. S., Sun, Y. M., and White, J. M. (1984) Coadsorption of CO and K on Ni(100). *Surf. Sci.* **141**, 82–100.
16. Bonzel, H. P., Broden, G., and Hrebs, H. J. (1983) X-ray photoemission spectroscopy of potassium promoted Fe and Pt surfaces. *Appl. Surf. Sci.* **16**, 373–394.
17. Kisinova, M., Pirug, G., and Bonzel, H. P. (1983) Coadsorption of potassium and CO on Pt(III). *Surf. Sci.* **133**, 321–343.
18. Brundle, C. R. (1976) XPS and UPS studies of interaction of nitrogen containing molecules with nickel. Use of binding energy patterns and relative intensities to diagnose surface species. *J. Vac. Sci. Technol.* **13**, 301–309.
19. Carley, A. F., Rassias, S., Roberts, M. W., and Wang, T. H. (1979) A study of the interaction of nitric oxide with nickel oxidized surfaces by X-ray photoelectron spectroscopy. *J. Catal.* **60**, 385–393.
20. Kishinova, M., Pirug, G., and Bonzel, H. P. (1984) NO adsorption on Pt(III). *Surf. Sci.* **136**, 285–295.
21. Umbach, E., Kulkarni, S., Feulner, P., and Menzel, D. (1979) A multimethod study of the adsorption of NO on Ru(001). I. XPS, UPS and ZAES measurements. *Surf. Sci.* **88**, 65–94.
22. Egawa, C., Naito, S., and Tamaru, K. (1984) Adsorption and decomposition of NO and NH_3 on Ru(001) and Ru(1, 1, 10) surfaces. *Surf. Sci.* **138**, 279–291.
23. Zhdan, P. A., Boreskov, G. K., Egelhoff, W. F., and Wainberg, W. N. (1976) An XPS investigation of the decomposition of NO on the Ir(III) surface. *J. Catal.* **45**, 281–290.
24. Fukuda, Y. and Rabalais, J. W. (1982) Chemisorption of NO and NH_3 on cobalt. *J. Electron Spectrosc. Rel. Phen.* **25**, 237–244.
25. Zhdan, P. A., Boreskov, G. K., Baronin, A. I. *et al.* (1977) Carbon monoxide oxidation by nitric oxide on iridium(III) studied by X-ray and u.v.-photoelectron spectroscopy. *Appl. Surf. Sci.* **1**, 25–32.
26. Joyner, R. W. and Roberts, M. W. (1978) Oxygen (1s) binding energies. *Chem. Phys. Lett.* **28**, 246–248.

27. Brundle, C. R. (1974) Discussion. *Faraday Discuss. Chem. Soc.* **58**, 131–132.
28. Haber, J., Stoch, J., and Unger, L. (1976) X-ray photoelectron spectra of oxygen in oxides of Co, Ni, Fe, and Zn. *J. Electron Spectrosc. Rel. Phen.* **9**, 459–468.
29. Schulze, P. D., Shaffer, S. L. *et al.* Adsorption of water on rhenium studied by XPS. *J. Vac. Sci. Technl.* **A1**, 97–99.
30. Wandelt, K. (1982) Photoemission studies of adsorbed oxygen and oxide layers. *Surf. Sci. Rep.* **2**, 1–121.
31. Brundle, C. R. and Bickley, R. I. (1979) Photoemission and Auger spectroscopy studies of the interaction of oxygen with zinc. *Faraday Trans. II* **75**, 1030–1046.
32. Moyes, R. B. and Roberts, M. W. (1977) Interaction of cobalt with oxygen, water vapor and carbon monoxide. X-ray and ultraviolet photoemission studies. *J. Catal.* **49**, 216–224.
33. Joyner, R. W. and Roberts, M. W. (1979) A study of the adsorption of oxygen on silver at high pressure by electron spectroscopy. *Chem. Phys. Lett.* **60**, 459–462.
34. Behndorf, C., Franck, M., and Thiewe, F. (1983) Oxygen adsorption on Ay(III) in the temperature range 100–500 K. *Surf. Sci.* **128**, 417–423.
35. Batra, I. P. and Kleinman, L. (1984) Chemisorption of oxygen on aluminium surfaces. *J. Electron Spectrosc.*, 1984, vol. 33, N3, p. 175–242.
36. Su C. Y., Lindau I., Chye P. W. *et al.* (1984) Photoemission studies of clean and oxidized Cs. *J. Electron Spectrosc. Rel. Phen.* **31**, 221–259.
37. Tatarenko, S., Dolle, P., Morancho, R. *et al.* (1983) XPS and UPS study of oxygen adsorption on Re(0001) at low temperature. *Surf. Sci.* **134**, L505–L512.
38. Brundle, C. R. and Carley, A. F. (1975) Oxygen adsorption in nickel surface detection of different species by X-ray photoelectron spectroscopy. *Chem. Phys. Lett.* **31**, 423–427.
39. Brundle, C. R. and Hopster, H. (1981) The kinetics and mechanism of the interaction of oxygen with Ni(100). *J. Vac. Sci. Technol.* **18**, 663–664.
40. Norton, P. R., Tapping, R. L., and Goodale, J. W. (1977) A photoemission study of the interaction of nickel (100), (110) and (111) surface with oxygen. *Surf. Sci.* **65**, 13–36.
41. Fuggle, J. C. and Menzel, D. (1975) Coverage dependent shifts of XPS peaks during chemisorption on metals. *Surf. Sci.* **53**, 21–34.
42. Barrie, A. and Bradshow, A. M. (1975) XPS shifts or substrate core levels upon chemisorption. *Phys. Lett.* **55A**, 306–308.
43. Miller, J. N., Ling, D. T., Shek, M. L. *et al.* (1980) Photoemission studies of clean and oxygen covered Pt 6 (111) × (100). *Surf. Sci.* **94**, 16–28.
44. Ferrer, S. and Somorjaj, G. A. (1980) UPS and XPS studies of the chemisorption of O_2, H_2 and H_2O on reduced and stoichiometric $SrTiO_3$ (111) surface; the effects of illumination. *Surf. Sci.* **94**, 41–46.
45. Demuth, J. E. (1980) The reaction of acetylene with Ni(100) and Ni(110) surfaces at room temperature. *Surf. Sci.* **93**, 127–144.
46. Zhdan, P. A., Boreskov, G. K., Boronin, A. I. *et al.* (1979) Nitric oxide adsorption and decomposition of the (111) and (110) surfaces of iridium. *J. Catal.* **60**, 93–99.
47. Eberhardt, W. and Kunz, C. (1978) Oxidation of Al single crystal surfaces by exposure to O_2 and H_2O. *Surf. Sci.* **75**, 709–720.
48. Broden, G., Rhodin, T. N., Brucker, C. *et al.* (1976) Synchrotron radiation study of chemisorptive bonding of CO on transition metals: polarization effect on Ir(100). *Surf. Sci.* **59**, 593–611.
49. Ducros, R., Alnot, H., Ehrhardt, J. J. *et al.* (1980) A study of the adsorption of several oxygen-containing molecules (O_2, CO, NO, H_2O) on Re(0001) by XPS, UPS and temperature programmed desorption. *Surf. Sci.* **94**, 154–168.

50. Park, Y. O., Masel, R. I., and Stolt, K. (1983) An XPS study of carbon monoxide and nitric oxide adsorption on Pt(410). *Surf. Sci.* **131**, L385–L386.
51. Bradshaw, A. M., Cederbaum, L. S., and Domcke, W. (1975) Ultraviolet photoelectron spectroscopy of gases adsorbed on metal surface. *Struct. Bond. (Berlin)* **24**, 133–169.
52. Yates, J. T. and Erickson, N. E. (1974) X-ray photoelectron spectroscopic study of the physical adsorption of xenon and the chemisorption of oxygen on tungsten. *Surf. Sci.* **44**, 489–514.
53. Yu, K. Y., McMenanin, J. C., and Spicer, W. E. (1975) Photoelectron spectra of condensed gases on an inert substrate. *J. Vac. Sci. Technol.* **12**, 286.
54. Brundle, C. R. and Carley, A. F. (1975) XPS and UPS studies of the adsorption of small molecules on polycrystalline Ni films. *Faraday Discuss. Chem. Soc.* **60**, 51–70.
55. Demuth, J. E. and Eastman, D. E. (1974) Photoemission observations of π–d bonding and surface-reactions of adsorbed hydrocarbons on Ni(111). *Phys. Rev. Lett.* **32**, 1123–1127.
56. Brucker, C. F. and Rhodin, T. N. (1979) Chemisorption bonding on metal surface using photoelectron spectroscopy. *Surf. Sci.* **86**, 638–648.
57. Klauber, C. and Buner, B. G. (1982) Comments on the paper. *Surf. Sci.* **121**, L513–L521.
58. Nefedov, V. I. (1980) Carbonyl structure and comparison of donor–acceptor characteristics of CO and related ligands. *Koord. Khim.* **6**, 163–214 (in Russian).
59. Khelmer, B. Y., Nefedov, V. I., and Mazalov, L. N. (1976) X-ray and X-ray photoelectron spectra of free and coordinated ethylene and benzol molecules. *Izv. Akad. Nauk SSSR Ser. Fiz.* **40**, 329–332 (in Russian).
60. Bare, S. R., Griffiths, K., Hoffmann, P. *et al.* (1982) A synchrotron radiation study of the electronic and geometric structure of CO on Pt(110). *Surf. Sci.* **120**, 367–375.
61. Fock, J.-H., Schmidt-May, J., and Koch, E. E. (1984) Electronic structure of solid pyridine and pyridine adsorbed on polycrystalline silver. *J. Electron Spectrosc. Rel. Phen.* **34**, 225–234.
62. Yu, K. Y., Spicer, W. E., Lindau, I. *et al.* (1976) Relationship of heat of chemisorption to π_1-level and σ-level shifts, a measured by photoemission. *J. Vac. Sci. Technol.* **13**, 277–279.
63. Schmeisser, D. and Jacobi, K. (1979) UV photoelectron spectroscopy of matrix-isolated and condensed CO and N_2. *Chem. Phys. Lett.* **62**, 51.
64. Dunlap, B. I. and Gadzuk, J. W. (1980) Surface plasmon relaxation energies for CO adsorbed on jellium. *Surf. Sci.* **94**, 89–104.
65. Jacobi, K. and Rotermund, H. H. UV photoemission from physicsorbed atoms and molecules. *Surf. Sci.* **116**, 435–455.
66. Wandelt, K. (1984) Surface characterization by photoemission of adsorbed xenon. *J. Vac. Sci. Technol.* **A2**, 802–807.
67. Demuth, J. E. and Schell–Sorokin, A. J. (1984) Rare gas titrative studies of Si(111) surfaces. *J. Vac. Sci. Technol.* **A2**, 808–811.
68. Ling, D. T. and Spicer, W. E. (1980) UPS and TDS studies of the adsorption of CO and H_2 on Cu/Ni. *Surf. Sci.* **94**, 403–423.
69. Shek, M. L., Stephan, P. M., Dindau, I., and Spicer, W. E. (1983) CO chemisorption in PtCu surfaces. *Surf. Sci.* **1**, 399–426, 427–437, 438–448.
70. Di Cenzo, S. B., Wertheim G. K., and Buchanan, D. N. E. (1982) XPS studies of adatom–adatom interactions. *Surf. Sci.* **121**, 411–420.
71. Bagus, P. S., Hermann, K., and Seel, M. (1981) Bonding and photoemission of chemisorbed molecules. *J. Vac. Sci. Technol.* **18**, 435–452.

72. Hermann, K. and Bagus, P. S. (1977) Binding and energy-level shifts of carbon monoxide adsorbed on nickel model studies. *Phys. Rev.* **B16**, 4195–4208.
73. Yu, H. L. (1978) Theoretical study of Co chemisorption on nickel and copper surfaces. *J. Chem. Phys.* **69**, 1755–1763.
74. Hoogewijs, R. and Vennic, J. (1979) Core hole electronic relaxation in chemisorption systems: a cluster model analysis of the charge transfer. *Solid State Commun.* **31**, 531–538.
75. Williams, A. R. and Lang, N. D. (1978) Core-level binding energy shifts in metals. *Phys. Rev. Lett.* **40**, 954–957.
76. Martensson, N. and Johansson, B. (1979) Core level binding shifts between free and condensed atoms. *Solid State Commun.* **32**, 791–794.
77. Finster, J. (1976) Die Photoelectronenspectroskopie und der Ubergang vom freien Atom zum Festkorper. *Wiss. Z. Karl-Marx-Univ. Leipzig* **4**, 449–466.
78. Watson, R. E., Perlman, M. L., and Herbst, J. F. (1976) Core level shifts in 3d transition metals and tin. *Phys. Rev.* **B13**, 2358–2365.
79. Ley, L., Kowalzyk, S. P., McFeely, F. R. *et al.* (1973) X-ray photoemission from zinc. Evidence for extra-atomic relaxation via semilocalized excitons. *Phys. Rev.* **B8**, 2392–2402.
80. Davis, D. W. and Shirley, D. A. (1972) A relaxation to core-level binding energy shifts in small molecules. *Chem. Phys. Lett.* **15**, 185–190.
81. Himpsel, F. J., Schwenther, N., and Koch, E. E. (1975) Ultraviolet photoemission spectroscopy of solid nitrogen and oxygen. *Phys. Stat. Solidi* **B71**, 615–621.
82. Citrin, P. H. and Hamann, D. R. (1974) Measurement and calculation of polarization and potential energy effects on core-electron binding energies in solid. X-ray photoemission of raregases implanted in noble metals. *Phys. Rev.* **B10**, 4948–4963.
83. Gibbs, R. A., Winograd, N., and Young, V. Y. (1980) X-ray photoemission studies of atom implanted matrices: Ni in carbon. *J. Chem. Phys.* **72**, 4799–4804.
84. Young, V. Y., Gibbs, R. A., and Winograd, N. (1979) X-ray photoemission studies of atom implanted matrices; Cu, Ag and Au in SiO_2. *Chem. Phys.* **70**, 5714–5721.
85. Gadzuk, J. W. and Doniach, S. A. (1978) Soluble relaxation model for core level spectroscopy on adsorbed atoms. *Surf. Sci.* **77**, 427–448.
86. Lang, N. D. and Williams, A. R. (1977) Core holes in chemisorbed atoms. *Phys. Rev.* **B16**, 2408–2419.
87. Bradshaw, A. M., Domcke, W., and Cederbaum, L. S. (1977) Intrinsic and extrinsic plasmon coupling in X-ray photoemission from core states of adsorbed atoms. *Phys. Rev.* **B16**, 1480–1488.
88. Gunnarsson, O. and Schönhammer, K. (1978) Plasmon effects on core level spectra of adsorbates. *Solid State Commun.* **26**, 147–150.
89. Hussain, S. S. and Newns, D. M. (1978) Plasmon versus shake-up satellites in XPS spectroscopy of adsorbed species. *Solid State Commun.* **25**, 1049–1052.
90. Gumhalter, B. (1979) Shake-up and relaxation effects in core spectra of adsorbates. *Surf. Sci.* **80**, 459–470.
91. Heskett, D., Plummer, E. W., and Messmer, R. P. (1984) A correlation between anomalous electronic and vibrational properties of chemisorbed molecules. *Surf. Sci.* **139**, 558–568.
92. Umbach, E. (1982) Satellite structures in photoemission spectra from different types of adsorbates. *Surf. Sci.* **117**, 482–502.
93. Kambe, K. (1979) Analysis of the electronic state of adsorbates on single crystals by angle-resolved photoelectron spectroscopy *Surf. Sci.* **86**, 620–630.

94. Liebsch, A. (1978) Photoemission and electronic properties of surface. J. Wiley, Chichester pp. 167–192.
95. Hermannson, J. (1977) Final-state symmetry and polarization effects in angle-resolved photoemission spectroscopy. *Solid State Commun.* **22**, 9–12.
96. Scheffler, M., Kambe, K., and Forstmann, F. (1978) Angle-resolved photoemission from adsorbates: theoretical considerations of polarization effects and symmetry. *Solid State Commun.* **25**, 93–100.
97. Jacobi, K., Scheffler, M., Kambe, K., and Forstmann, F. (1977) Angle-resolved photoemission of the p(2 × 2) oxygen overlayer on Ni(001): measurements and calculations. *Solid State Commun.* **22**, 17–20.
98. Nyberg, G. L. and Richardson, N. V. (1979) Symmetry analysis of angle-resolved photoemission polarization dependence and lateral interaction in chemisorbed benzene. *Surf. Sci.* **85**, 335–352.
99. Richardson, N. V., Lloyd, D. R., and Quinn, C. M. (1979) Application of simple selection rules to photoemission from clean and adsorbate covered metal single crystals. *J. Electron Spectrosc. Rel. Phen.* **15**, 177–189.
100. Horn, K., Bradshaw, A. M., and Jakobi, K. (1978) Angular-resolved UV photoemission from ordered layers of carbon monoxide on nickel(100) surface. *Surf. Sci.* **72**, 719–732.
101. Horn, K., Bradshaw, A. M., and Jakobi, K. (1978) Studies of chemisorption of ethylene and acetylene on nickel(100). *J. Vac. Sci. Technol.* **15**, 575–579.
102. Shirley, D. A., Stöhr, J., Wehner, P. S. *et al.* (1977) Photoemission from noble metals and adsorbates using synchrotron radiation. *Phys. Scr.* **16**, 398–413.
103. Apai, G., Wehner, P. S., Williams, R. S. *et al.* (1976) Orientation of CO on Pt(III) and Ni(III) surface from angle-resolved photoemission. *Phys. Rev. Lett.* **37**, 1497–1500.
104. Smith, R. J., Anderson, J., and Lapeyre, G. J. (1976) Adsorbate orientation using angle-resolved polarization-dependent photoemission. *Phys. Rev. Lett.* **37**, 1081–1084.
105. Davenport, J. W. (1976) Ultraviolet photoionization cross-sections for N_2 and CO. *Phys. Rev. Lett.* **36**, 945–949.
106. Williams, P. M., Butcher, P., Wood, J., and Jakobi, K. (1976) Angle-resolved photoemission studies of nickel (111) surface and interaction with carbon monoxide. *Phys. Rev.* **B14**, 3215–3226.
107. Allyn, C. L., Gustafsson, T., and Plummer, E. W. (1977) The orientation of CO adsorbed on Ni(100). *Chem. Phys. Lett.* **47**, 127–132.
108. Allyn, C. L., Gustafsson, T., and Plummer, E. W. (1977) The chemisorption of CO on Cu(100) studied with angle-resolved photoelectron spectroscopy. *Solid State Commun.* **24**, 531–537.
109. Gustafsson, T. (1980) On the energy dependence of the angular distribution of photoelectrons from oriented molecules. *Surf. Sci.* **94**, 593–614.
110. Rieger, D., Schnell, R.D., and Steinmann, W. (1984) Angular distribution patterns of photoelectrons from orbitals of CO adsorbed on Ni(100), Pt(111) and Pt(110). *Surf. Sci.* **143**, 157–176.
111. Trenary, M., Tang, S. L. *et al.* (1983) An angle-resolved photoemission study of Co chemisorbed on the Pt(111) surface. *Surf. Sci.* **124**, 555–562.
112. Miranda, R., Wandelt, K., Rieger, D., and Schnell, R. D. (1984) Angle-resolved photoemission of CO Chemisorption on Pd(111). *Surf. Sci.* **139**, 430–442.
113. Engelhoff, W. F. (1984) N_2 on Ni(100). *Surf. Sci.* **141**, L324–L328.
114. Lloyd, D. R., Quinn, C. M., and Richardson, N. V. (1977) The angular distribution of

electrons photoemitted from benzene molecules adsorbed on nickel and palladium single crystal surfaces. *Solid State Commun.* **23**, 141–145.
115. Smith, N. V., Larsen, P. K., and Chiang, S. (1977) Anisotropy of core-level photoemission from InSe, gase and cesiated W(001). *Phys. Rev.* **16**, 2699–2706.
116. Fadley, C. S., Kono, S., Petersson, L. G. *et al.* (1979) Determination of surface geometries from angular distribution of deep-level X-ray photoelectrons. *Surf. Sci.* **89**, 52.
117. Norman, D., Farrell, H. H., Traum, M. M. *et al.* (1979) Photoelectron diffraction observations of adsorbates on nickel surfaces. *Surf. Sci.* **89**, 51.
118. Liebsch, A. (1976) Theory of photoemission from localized adsorbate levels. *Phys. Rev.* **13**, 544–555.
119. Li, C. H. and Tong, S. Y. (1979) Selective structural sensitivity and simplified computation of angle-resolved ultraviolet photoemission spectroscopy. *Phys. Rev. Lett.* **43**, 526–529.
120. Thompson, K. A. and Fadley, C. S. (1984) X-ray photoelectron diffraction from adsorbate core levels in the energy range 500–10 000 eV and with polarized radiation. *J. Electron Spectrosc. Rel. Phen.* **33**, 29–50.
121. Orders, P. J., Kono, S., Fadley, C. S. *et al.* (1982) Angle-resolved X-ray photoemission from core levels of C(2 × 2)Co on Ni(001). *Surf. Sci.* **119**, 371–383.
122. Lindgren, S. A. and Wallden, L. (1979) Adsorbate-induced angle averaging of Cu(111) photoemission spectra. *Phys. Rev. Lett.* **43**, 460–463.
123. Baston, J. J., Bahr, C. C., Hussain, Z. *et al.* (1984) Direct determination of surface structures from photoelectron diffraction. *J. Vac. Sci. Technol.* **A2**, 847–851.
124. Schlüter, M., Rowe, J. E., Margaritondo, G. *et al.* (1976) Chemisorption-site geometry from polarized photoemission: Si(111)Cl and Ge(111)Cl. *Phys. Rev. Lett.* **37**, 1632–1635.
125. Rowe, J. E., Margaritondo, G., and Christman, S. B. (1977) Chlorine chemisorption on silicon and germanium surface. Photoemission polarization effects with synchrotron radiation. *Phys. Rev.* **B16**, 1581–1589.
126. Schlüter, M., Rowe, J. E., and Weeks, S. P. (1979) Chemisorption of Cl in surface vacancies on Si(111)1 × 1. *J. Vac. Sci. Technol.* **16**, 615–617.
127. Larsen, P. K., Smith, N. V., Schlüter, M. S. *et al.* (1978) Surface energy bonds and atomic position of Cl chemisorbed on cleaved Si(111). *Phys. Rev.* **B17**, 2612–2619.
128. Menon, P. G. and Prasada Rao, T. S. R. (1979) Surface enrichment in catalysts. *Catal. Rev. Sci. Eng.* **20**, 97–120.
129. Grzybowski, B., Haber, J., Marczewalski, W., and Ungier, L. (1978) X-ray and ultraviolet photoelectron spectra of bismuth molybdate catalysts. *J. Catal.* **42**, 327–333.
130. Prasada Rao, T. S. R. and Menon, P. G. (1978) Physicochemical studies on silica-supported multicomponent molybate catalyst before and after use ammoxidation of propyiene. *J. Catal.* **51**, 64–71.
131. Brinen, J. S. (1974) Application of ESCA to industrial chemistry. *J. Electron Spectrosc. Rel. Phen.* **5**, 377–383.
132. Haber, J., Nowotny, J., Sikora, I., and Stoch, J. (1984) Electron spectroscopy in studies of surface segregation of Cr in Cu-doped CoO. *Appl. Surf. Sci.* **17**, 324–330.
133. Carbossi, F. and Petrini, G. (1984) XPS study on the low-temperature CO shift reaction catalyst. *J. Catal.* **90**, 106–112, 113–118.
134. Duquette, L. G., Cieslinski, R. C., Jung, C. W. *et al.* (1984) ESCA studies on silica and aluminia-supported rhenium oxide catalysts. *J. Catal.* **90**, 362–365.
135. Muralidhar, G., Concha, B. E., Grey, L. *et al.* (1984) Characterization of reduced and sulfided supported molybdenum catalysts. *J. Catal.* **89**, 274–284.

136. Markovsky, L. E., Stencel, J. M., and Brown, F. R. (1984) A surface spectroscopic study of Cu–Mo/Al_2O_3 catalysts. *J. Catal.* **89**, 334–347.
137. Sexton, B. A., Hughes, A. E., and Foyer, K. (1984) An XPS and reaction study of Pt–Sn catalysts. *J. Catal.* **88**, 466–477.
138. Dang, T. A., Petrakis, L., Kibby, C., and Hercules, D. H. (1984) Surface characterization of Th–Ni intermetallic catalysts. *J. Catal.* **88**, 26–36.
139. Shulga, Y. M., Ivleva, I. N., Shimanskaya, M. V. *et al.* (1975) Electron structure and surface composition of V_2O_5–MoO_3 heterogeneous oxidation catalyst. *Zh. Fiz. Khim.* **49**, 2976–2977 (in Russian).
140. Shulga, Y. M., Karklin, L. N., Shimanskaya, M. V. *et al.* (1977) X-ray photoelectron spectra of vanadium–molybdenum catalysts modified by metal oxides. *Zh. Fiz. Khim.* **51**, 1234–1235 (in Russian).
141. Matsuura, I. and Wolfs, M. W. J. (1975) X-ray photoelectron spectroscopy study of some bismuth molybdates and multicomponent molybdates. *J. Catal.* **37**, 174–178.
142. Minachev, Kh. M., Khodakov, Y. S., Antoshin, G. V. *et al.* (1978) Study of thermal activation and deactivation of alumochrome catalysts. *Izv. Akad. Nauk SSSR Ser. Khim.* 549–551 (in Russian).
143. Slinkin, A. A., Antoshin, G. V., Loktev, M. N. *et al.* (1978) On the phase composition and nature of reduction of VaO–Al_2O_3, MoO_3–Al_2O_3, CoO–MoO_3–Al_2O_3 catalysts. *Izv. Akad. Nauk SSSR Ser. Khim.* 2225–2233 (in Russian).
144. Lorenz, P., Finster, J., Wendt, G. *et al.* (1979) ESCA investigations of some NiO/SiO_2 and NiO–Al_2O_3 SiO_2 catalysts. *J. Electron Spectrosc. Rel. Phen.* **16**, 267–276.
145. Okamoto, Y., Shimokawa, T., Imanaka, T., and Teranishi, S. (1979) X-ray photoelectron study of Co–Mo binary oxide catalysts. *J. Catal.* **57**, 153–166.
146. Richter, K., Peplinski, B., Hebisch, H., and Kleinschmidt, G. (1980) Study on aging of multicomponent catalyst of SiO_2 support in long term test using X-ray photoelectron spectroscopy. *Appl. Surf. Sci.* **4**, 205–213.
147. Andersson, S. T. and Järås, S. (1980) Activity measurement and ESCA investigation of a V_2O_5/SnO_2 catalyst for the vapor-phase oxidation of alkylpyridines. *J. Catal.* **64**, 51–67.
148. Edmonds, T. and Mitchell, P. C. H. (1980) The XPS of some MoO_3/Al_2O_3 catalysts and the distribution of molybdenum in catalysts extrudates following drying and calcining. *J. Catal.* **64**, 491–493.
149. Gajardo, P., Grange, P., and Delmon, B. (1980) Structure of oxide CoMo/Al_2O_3 hydrodesulphurization catalysts: an XPS and DRS study. *J. Catal.* **63**, 201–216.
150. Gajardo, P., Pirotte, D., Defosse, C. *et al.* (1979) XPS study of the supported phase SiO_2 interaction in Mo/SiO_2 and CoMo/SiO_2 hydrodesulphurization catalysts in their oxidic precursor form. *J. Electron Spectrosc. Rel. Phen.* **17**, 121–135.
151. Best, S. A., Squires, R. G., and Walton, R. A. (1977) X-ray photoelectron spectra of heterogeneous catalysts chromica–silica catalyst system. *J. Catal.* **47**, 292–299.
152. Cimino, A., Angelis, B. A., de, Luchetti, A., and Minelly, G. (1976) Characterization of CROX-S 102 catalysts by photoelectron spectroscopy; X-ray and optical measurements. *J. Catal.* **45**, 316–325.
153. Finster, J., Lorenz, P., and Meisel, A. (1979) Quantitative ESCA surface analysis applied to catalysts: investigation of concentration gradients. *Surf. Interface Anal.* **1**, 179–184.
154. Cimino, A. and Minelly, G. (1978) Investigation of the surface compositions by X-ray photoelectron spectroscopy. *J. Electron Spectrosc. Rel. Phen.* **13**, 291–303.
155. Seah, M. P. (1979) Surface science in metallurgy. *Surf. Sci.* **80**, 8–23.

156. Overburg, S. H., Bertrand, P. A., and Somorjai, G. A. (1975) Surface composition of binary systems, prediction of surface phase-diagrams of solid solutions. *Chem. Rev.* **75**, 547–560.
157. Kelley, M. (1979) Surface segregation: a comparison of models and results. *J. Catal.* **57**, 113–125.
158. Sachtler, W. M. H. and Van Santen, R. A. (1979) Surface composition of binary alloys. *Appl. Surf. Sci.* **3**, 121–144.
159. Kelley, M. J. and Ponec, V. (1982) Surface composition of alloys. *Progr. Surf. Sci.* **11**, 139–244.
160. Zhdan, P. A. and Boreskov, G. K. (1975) Study of oxide copper-magnesium catalysts by X-ray photoelectron spectroscopy. *Dokl. Akad. Nauk SSSR* **224**, 1348–1351 (in Russian).
161. Knecht, J. and Stork, G. (1977) Quantitative analysis of zeolites with X-ray photoelectron spectroscopy; pure and copper-doped zeolites. *Z. Anal. Chem.* **283**, 105–108.
162. Anderson, S. L. T. and Scurrell, M. S. (1979) Infrared and ESCA studies of a heterogenized rhodium carbonylation catalyst. *J. Catal.* **59**, 340–356.
163. Knecht, J. and Stork, G. (1977) Quantitative analysis of zeolites with X-ray photoelectron spectroscopy; iron-doped zeolites. *Z. Anal. Chem.* **286**, 47–49.
164. Tempere, J. F., Delafosse, D., and Contour, J. P. (1975) X-ray photoelectron spectroscopy study of zeolites. *Chem. Phys. Lett.* **33**, 95–98.
165. Primet, M., Vedrine, J. C., and Naccache, C. (1978) Formation of rhodium carbonylcomplexes in zeolite. *J. Mol. Catal.* **4**, 411–421.
166. Finster, J. and Lorenz, P. (1977) Photoelectron intensity spectroscopy: analytical and structural information from PEIS applied to zeolites. *Chem. Phys. Lett.* **50**, 223–227.
167. Finster, J., Lorenz, P., and Angele, E. (1978) Photoelektronen spektroskopische Untersuchungen an AgNaA bzw AgCaNaA-Zeolithen. *Z. Phys. Chem.* **259**, 113–122.
168. Vedrine, J. C., Dufaux, M., Nassache, C., and Imelik, B. (1978) Photoelectron spectroscopy study of ions in type Y-zeolite. *J. Chem. Soc. Faraday Trans. I* **74**, 440–449.
169. Kuznicki, S. M. and Eyring, E. M. (1980) An ESCA study of Rh(III) exchanged zeolite catalysts. *J. Catal.* **65**, 227–230.
170. Pedersen, L. A. and Lunsford, J. H. (1980) A study of ruthenium in zeolite-Y by X-ray photoelectron spectroscopy. *J. Catal.* **61**, 39–47.
171. Tkachenko, O. P., Shpiro, Y. S., Antoshin, G. V., and Minachev, Kh. M. (1980) State of ruthenium in zeolite-V and its catalytical activity in reaction of NO-reduction. *Izv. Akad. Nauk. SSSR Ser. Khim.* **6**, 1249–1256 (in Russian).
172. Barr, T. L. (1983) An XPS study of Si as it occurs in adsorbents, catalysts and thin films. *Appl. Surf. Sci.* **15**, 1–15.
173. Givens, K. E. and Dillard, J. G. (1984) Hydrodesulfurization of thiophene using Rn(III) zeolites. *J. Catal.* **86**, 108–120.
174. Minachev, Kh. M., Shpiro, Y. S., Tkachenko, O. P., and Antoshin, G. V. (1984) Study of zeolite catalysts formation with the use of some physical methods of surface analysis. *Izv. Akad. Nauk. SSSR Ser. Khim.* 3–26 (in Russian).
175. Slinkin, A. A., Antoshin, G. V., Loktev, M. I. *et al.* (1978) On the migration of cations during the oxidation–reduction processing of CuNiY zeolite. *Kin. Katal.* **19**, 754–759 (in Russian).
176. Riassian, M., Trimm, D. L., and Williams, P. M. (1976) Support effects in break-up and aggregation of silver films under catalytic conditions. *J. Chem. Soc. Faraday Trans. I* **1**, 926–929.

177. Hoste, S., Vondel, D. V., and Kelen, G. P. V. D. (1979) XPS of a steam reforming NiAl 204 catalyst. *J. Electron Spectrosc. Rel. Phen.* **16**, 407–413.
178. Carberry, J. J., Kuczynski, G., and Martinez, E. (1972) Influence of gammairradiation upon catalytic selectivity. *J. Catal.* **26**, 247–252.
179. Brinen, J. C., Schmitt, J. L., Doughman, E. W. *et al.* (1975) X-ray photoelectron study of rhodium on charcoal catalyst. *J. Catal.* **40**, 295–300.
180. Brinen, J. C. and Schmitt, J. L. (1976) Electron spectroscopy for chemical analysis intensity ratio-probe of catalyst structure. *J. Catal.* **45**, 274–276.
181. Agevine, P. J., Deglass, W. N., and Vartuli, J. C. (1976) Dispersion and uniformity of supported catalysts by X-ray photoelectron spectroscopy. Proceedings of the Sixth International Congress on Catalysis, Vol. 2, pp. 611–620.
182. Escard, J., Poutvianne, B., Chenebaux, M. T., and Cosyns, J. (1976) ESCA characterization of noble metal based catalysts on supports; study of platinum-based catalysts deposited on alumina. *Bull. Soc. Chim. France* **N3/4**, 349–354.
183. McIntyre, N. S., Sagert, N. H., Pouten, P. M. L., and Proctor, W. G. (1973) Surface analysis of nickel chromia catalysts by X-ray photoelectron spectroscopy. *Can. J. Chem.* **51**, 1670–1672.
184. Briggs, D. (1976) ESCA and metal crystallite size/dispersion in catalysts. *J. Electron Spectrosc. Rel. Phen.* **9**, 487–491.
185. Fung, S. C. (1979) Application of XPS to determination of the supported particles in a catalyst: Model development and its application to describe the sintering behaviour of a silica-supported Pt film. *J. Catal.* **58**, 454–468.
186. Houalla, M., Delanay, F., and Delmon, B. (1982) The dispersion of nickel oxide supported on boron-modified silicas. *J. Electron Spectrosc. Rel. Phen.* **25**, 59–66.
187. Perevezentseva, N. N., Lebedenko, Y. B., Zhavoronkova, K. N., and Nefedov, V. I. (1984) X-ray photoelectron study of Ni distribution on the surface of various carriers. *Poverkhnost* **3**, 109–112 (in Russian).
188. Legare, P., Sakisaka, Y., Brucker, C. F., and Rhodin, T. N. (1984) Electronic structure of highly dispersed supported transition metal clusters. *Surf. Sci.* **139**, 316–332.
189. Houalla, M. and Delmon, B. (1981) Use of XPS to detect variations in dispersion of impregnated and ion-exchanged NiO/SiO_2 system. *Surf. Interface Anal.* **3**, 103–105.
190. Fleisch, T. H., Hick, R. F., and Bell, A. T. (1984) An XPS study of metal-support interactions on Rd/SiO_2 and Pd/La_2O_3. *J. Catal.* **87**, 398–413.
191. Huizinga, T., Van'T Blick, H. F. J., Vis, J. C., and Prins, R. (1983) XPS investigations of Pt and Rh supported on γ-Al_2O_3 and TiO_2. *Surf. Sci.* **135**, 580–596.
192. Cheang, T. T. P. (1984) X-ray photoemission of small platinum and palladium clusters. *Surf. Sci.* **140**, 151–154.
193. Wertheim, G. K., DiCenzo, S. B., and Youngquist, S. E. (1983) Unit charge on supported gold clusters in photoemission final state. *Phys. Rev. Lett.* **51**, 2310–2314.
194. Lunin, V. V., Nefedov, V. I., Zhumadilov, E. K. *et al.* (1978) Surface segregation influence on the catalytical activity of the Zr–Ni–H system. *Dokl. Akad. Nauk. SSSR* **240**, 114–116 (in Russian).
195. Lunin, V. V., Nefedov, V. I., Erivanskaya, L. A., and Rakhmimov, B. Y. (1980) Surface composition influence on catalytical properties of Zr–Cu–Ni in toluene hydromethylation. *Zh. Fiz. Khim.* **34**, 1853–1858 (in Russian).
196. Nefedov, V. I., Sergushin, N. P., and Ryashentseva, M. A. (1973) X-ray photoelectron study of rhenium sulphide catalysts. *Dokl. Akad. Nauk. SSSR* **213**, 600–603 (in Russian).
197. Ustinova, T. S., Shpiro, Y. S., Smirnov, V. S., *et al.* (1976) Changes in the composition

of the surface layer of palladium and rhodium compounds under the influence of a catalytical reaction. *Izv. Akad. Nauk. SSSR Ser. Khim.* **2**, 441–444 (in Russian).

198. Boudeville, Y., Fineras, F., Forissier, M. *et al.* (1979) Correlation between X-ray photoelectron spectroscopy data and catalytic properties in selective oxidation on Sb–Sn–O catalysts. *J. Catal.* **58**, 52–60.
199. Fasman, A. B., Kalina, M. M., Minachev, Kh. M. *et al.* (1979) Bulk and surface composition of rhenium copper catalysts. *Izv. Akad. Nauk. SSSR Ser. Khim.* 244–246 (in Russian).
200. Shpiro, Y. S., Ustinova, T. S., Smirnov, V. S. *et al.* (1978) Influence of processing nature on the surface layer composition of palladium binary alloys. *Izv. Akad. Nauk. SSSR Ser. Khim.* 763–767 (in Russian).
201. Contour, J. P., Mouvier, G., Hoogewys, M., and Hecleve, G. (1977) X-ray photoelectron spectroscopy and electron microscopy of Pt–Rh gauzes used for catalytic oxidation of ammonia. *J. Catal.* **48**, 217–228.
202. Busch, C., Schapbach, L., Stucki, F. *et al.* (1979) Hydrogen storage. Surface segregation and its catalytic effect on hydrogeneration and structural studies by means of neutron diffraction. *Int. J. Hydrogen Energy* **4**, 29–39.
203. Ponec, V. (1979) Surface composition and catalysis on alloys. *Surf. Sci.* **80**, 352–366.
204. Verbeck, B. H. (1982) Core-level shifts in metallic compounds. *Solid State Commun.* **44**, 951–953.

Chapter 5

Studies of the surface and processes on the surface

5.1. Surface electron states

The ultimate size of solids does not exert a substantial influence on a compound's electronic structure if the solid's linear dimensions are large compared to interatomic distance, i.e. if the number of border atoms is small as compared to the number of atoms inside the solid. In this sense it is customary to refer to the electronic structure of the substance as the electronic structure of the bulk. When such an electronic structure is calculated, the solid often is assumed to be infinite in order to make full use of the periodicity conditions of the geometrical lattice. But in all real samples there are atoms on the surface of the body for which several physical characteristics (for instance electrostatic potential, geometry of surrounding medium, etc.) sharply differ from those for atoms in bulk. It is also necessary to stress that often atoms in the surface layer change their mutual position relative to the bulk of the body. In fact, a multitude of rearrangements of the surface layer, depending on the nature of the surface and the monocrystal's overall structure, are dealt with in literature.*

One of the best known models [1] for (111) surfaces with a (2×2) structure of elementary semiconductors of the type Si, Ge, etc., and of A^3B^5 compounds, considers the alternate raising and lowering of (111) surface atoms. If we proceed from the sp^3-hybridization of these compounds, the free unsaturated valence of the atom ('dangling' bond) on the surface should accord with the sp^3 hybrid orbital, while the three other such orbitals ensure this atom's bond with the three neighbouring ones. But rehybridization should occur as a result of change in the geometric position of the atoms. Model [1] presupposes that the atoms that have descended deep into the lattice strive to form flat sp^2 bonds, while the 'dangling' (unsaturated) bond approaches the p_z-state. In contrast, the ascended atoms strive to form three bonds at right angle (p^3-hybridization), while the unsaturated bond approaches the s-state.

It follows from this simple example, just as from the most general considerations, that electronic structure on the border of the surface can differ substantially from the electronic structure in bulk. Moreover, one can discern levels corresponding to the atom's bonds with its neighbours, and also to unsaturated bonds. A classification of this type makes sense for solids of the type Si, GaAs, and others where it is possible to single out localized bonds between

*Very often interatomic distances in the surface layer diminish.

neighbouring atoms. Surface states cannot be thus classified for metals. Usually they are viewed as sub-zones split away from the solid's electronic zones as a result of changes of the potential on the surface. Such surface states (Tamm state) were recently discovered in X-ray photoelectron spectra of copper [2] and the Cu–Al alloy [3]. In the general case the electron surface states can be placed in the energy range of the valence band or in the energy gap of the solid's electronic states.

When analysing photoelectron and X-ray photoelectron spectra the following methods are often used to distinguish surface states from bulk ones. First, a comparison is made of the photoelectron spectra of the valence band of a clean surface with spectra after the adsorption of some gas capable of saturating the 'dangling' bonds, which should bring about a disappearance of the corresponding bands in the spectrum. Photoelectron He(I) spectra of the Si(100) (2 × 1) surface before and after the adsorption of H_2 are given as an example in Fig. (5.1) [4]. The

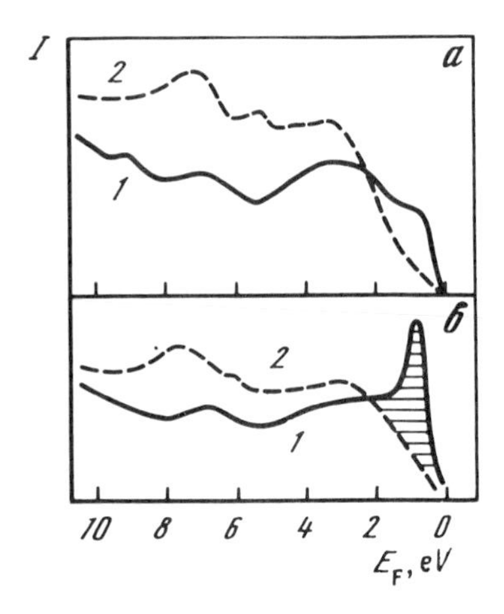

Figure 5.1. He(I) spectra of Si(100) (2 × 1) surface (1) before and (2) after adsorption of H_2.

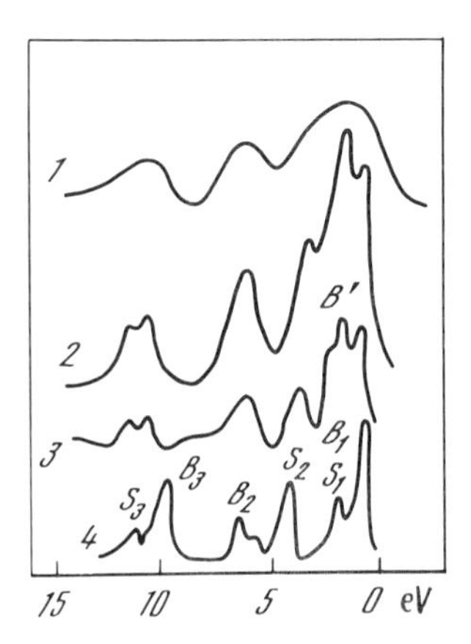

Figure 5.2. GaAs X-ray photoelectron spectra.

spectrum in Fig. (5.1a), represents the intensities integrated with respect to all angles, while the spectrum in (Fig. 5.1b), gives the intensity for the normal fall of radiation on the surface and the photoelectrons counting perpendicular to the beam and parallel to the polarization vector of ionizing radiation. The shaded part of the spectrum corresponds to surface states.

Second, a study is made of the X-ray photoelectron spectra of valence bands for two different photoelectron exit angles θ from the solid. As was explained in Chapter 2, spectra at small θ angles are more sensitive to the surface layer than spectra at large θ. On the basis of formulae cited in Chapter 2 it is possible to evaluate the relative contribution of intensity at large values of θ (bulk spectra) to the spectrum at small values of θ. The difference between spectra yields the contribution of surface states. As an example, Fig. (5.2) presents X-ray photoelectron spectra of GaAs(110) [5] at $\theta = 65^\circ$ (1) and 5° (2). Spectrum 3 corresponds to the difference of spectra 1 and 2, the intensity of spectrum 1 being multiplied by 0.3 (this is the relative contribution of the bulk state to spectrum 2). The curve 4 corresponds to the calculated density of the surface states.

The use of a roughly similar procedure enabled the authors of [6] not only to divide the valence band spectra of the first surface layer of Au and the rest of the sample but also to discover experimentally for the first time two components of the 4f inner level line of Au at extremely small photoelectron exit angles. The component with the lesser binding energy (0.40 ± 0.01 eV) corresponded to the surface layer (see Section 5.2).

At present surface states have been discovered on the basis of photoelectron data in metals (see, for instance, [2, 7, 8]) and semiconductors [5, 9–21]. Within the framework of the present section let us dwell on some results for Si and GaAs.

Reference [9] studies the surface state Si(111), which yields photoelectron spectrum maxima with an energy 1.15 eV below the Fermi level. If the polar angle θ is increased from 15 to 65° while maintaining a constant azimuthal angle $\Phi = 0^\circ$, a small decrease of this maximum energy is observed, while at $\theta > 45^\circ$ a certain splitting of the maximum becomes noticeable. A study of the azimuthal dependence of intensity revealed three intensity maxima for directions corresponding to the bisectors of the Si–Si bond angles for the atom under investigation. The conclusion was drawn on this basis that the said maximum cannot correspond to an unsaturated bond with p_z symmetry. An admixture of states responsible for the Si–Si bond between atoms in the crystal's first and second layers is presumed. An analysis of the angular dependence of Si(100) [4] surface state spectra showed that none of the three models of the structure of this surface existing in the literature explain the observed spectrum. A review of studies of 'dangling' bonds on the Si(111) surface on the basis of photoemission angular dependence is given in [20].

The angular dependence of the intensity of He(I) and Ne(I) photoelectron spectra on the surfaces GaAs $(\bar{1}\bar{1}\bar{1})$ (2×2) and $(\bar{1}\bar{1}\bar{1})$ $(\sqrt{19} \times \sqrt{19})$ is studied in reference [11]. Surface states with energies of 1.8, 3.0, 3.8, and 7.0 eV were discovered. The intensity of the state with an energy of 1.8 eV sharply increases during the transition from s-polarized radiation (the polarization vector of ionizing radiation is perpendicular to the escape direction of photoelectron, i.e.

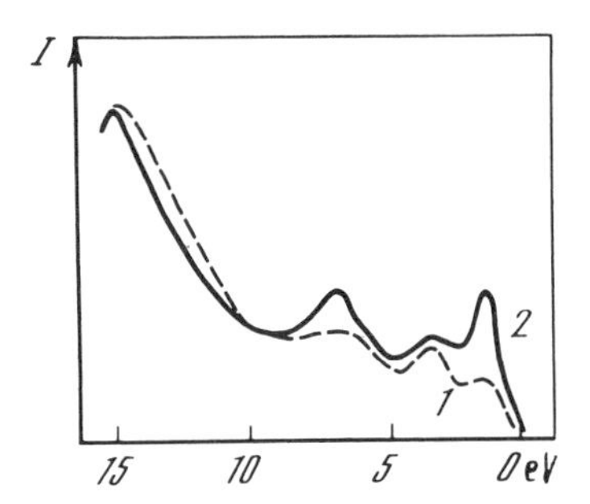

Figure 5.3. He(I) photoelectronic spectrum of GaAs(III) surface (2 × 2). Incidence angle of radiation = 45°; polar angle of photoelectron escape = 49°. 1 = s-polarization of He(I) radiation; 2 = p-polarization.

A ⊥ *xy*) to p-polarized (**A** ∥ *xy*) (Fig. 5.3). Since by the selection rules (see Chapter 4, Section 4.2) this should take place for a_1 symmetry levels, the maximum is referred to the unsaturated 'dangling' bonds of antimony atoms localized perpendicular to the plane. The intensity of this maximum noticeably decreases during the annealing of the surface, that is during the transition $(\bar{1}\bar{1}\bar{1})$ (2 × 2)→(111) ($\sqrt{19} \times \sqrt{19}$). This was interpreted as a change in the nature of the 'dangling' orbital: the transition from the p_z state to the s state, as it follows from the Haneman mode (see above). The photoionization cross-section of valence s-states of non-transitional elements is smaller than of p-states for the He(I) radiation, and it is this that brings about the decrease of the maximum's intensity during annealing. The surface states with energies of 4 and 7 eV are referred to the bonds of antimony atoms on the surface with the neighbouring Ga underlayer.

A number of references [5, 10, 12–15] are devoted to GaAs(110) surface states. Surface states, taking due account of three possible rearrangements of the surface are calculated in [15]. A comparison with experimental data of photoelectron spectra shows that the model of bond relaxation in which As(Ga) atoms in the first layer ascend (descend) by $0.1(0.25)d_0$ from their position in the solid (d_0 = 0.2 nm—interatomic distance) accords better than the rest with the experiment. In the second layer the atoms shift 0.05 nm to the side opposite to the shift of atoms of the same sort in the first monolayer. This model is analogous to the already considered Haneman model. A calculation [15] of electron density for this model [Fig. (5.2), curve 4] leads to maxima S_i and B_i. Maximum B corresponds to the 'dangling' unsaturated bonds—this is mostly state As $4p_z$. The maxima S_1 and S_2 also predominantly correspond to 4p-states of As (p_y and $p_{x,y}$ symmetry)—they correspond to the bonds between p-states of As and Ga, while the B_2 and B_3 maxima accord with the Ga 4s–As 4p–Ga 4p bonds correspondingly.

Maxima with energies of 1.1, 1.9, 3.7, 6.5, 11.0, and 12.0 eV, which are close to the data of photoelectron spectra [12] have been found in the X-ray photoelectron spectra [5] (Fig. 5.2). It is easy to interpret the spectra of surface states on the basis of Fig. (5.2). It shall be noted that in reference [13] the S_2 state is referred to the energy of 1.9 eV, this corresponding to the shoulder B' on the curve 3 in

Fig. (5.2) with an energy of 2.7 eV. This interpretation follows from an analysis of the angular dependence of intensity.

In conclusion it shall be noted that the interpretation of surface states, just as their identification itself, is a rather complex and not always uniform procedure. As an example let us dwell on the results of [14], in which it is shown that both the energy position and angular dependence of the GaAs(110) photoelectron spectrum intensity maxima, which are usually referred to surface states, can be explained with due regard only for electron structure characteristic of the GaAs bulk.

5.2. Binding energy differences between atoms in bulk and on the surface

Both the electron state and the Madelung potential are different for atoms in bulk and on surface. As a result, differences in E_b, the binding energy, should be expected. For the first time such a difference was reliably discovered for gold [6, 22]. The low-energy component, with a relative intensity that increased with the lessening of the angle between the electron escape and the surface, was found on the line Au $4f_{7/2}$. A component corresponding to Au atoms on the surface and having a binding energy 0.4 eV less than the line corresponding to Au atoms in bulk was singled out.

Also given in this reference [22] is an explanation of this negative shift of the lines of surface atoms on the basis of electron considerations, that was developed and specified in subsequent papers [23–25]. According to these considerations the direction of the lines shift of surface atoms is due to the decrease of the coordination number for surface atoms as compared to bulk. This leads to a decrease of the d-band width. The decrease is at the expense of the lessening of the delocalization of d-electrons and sp-hybridization, that is the number of localized d-states increases at the expense of d- and s-states. As a result of the narrowing of the d-band there takes place a change in the position of the Fermi level. Besides, in noble elements and in elements at the end of the periodic table where the electronic configuration is $d^n (n > 5)$, the Fermi level of surface atoms should shift in the direction of lower energies. But the Fermi level is held steady by the atoms in the bulk; for this reason the shift ΔE_{sb} occurs both for inner levels and for the mean position of the valence band ΔE towards lower binding energies for surface atoms. Experiment [25, 26] has, indeed, confirmed an 8% decrease in the valence band width of Au surface atoms, a shift of 0.5 eV as compared to these characteristics of bulk atoms. A rough equality of the values $\Delta E_{sb} = -0.24$ and $\Delta E = -0.4$ is observed also for Cu [26]. The coincidence of the signs ΔE_{sb} and ΔE is also true for Ir, Ta [23].

For elements at the start of the Periodic Table with the configuration $d^n (n < 5)$, the Fermi level of surface atoms shifts in the direction of greater energies, bringing about a positive line shift in the surface atoms. This, too, is confirmed by experiment (see review [25]).

Extensive experimental material has now been obtained on the shifts of the inner lines of surface atoms (see [22–43] and references in these papers). Some data are presented in Tables (5.1–5.3). ΔE_{sb} shifts have been found not only in

Table 5.1
Shifts of ΔE_{sb} surface levels as compared to bulk [33], eV

Metal	Line	E_{sb}(experiment)	E_{sb}(theory)
Au	4f	−0.39	−0.6
Pd	3d	0.23	−0.4
Rh	3d	−0.58	−0.6
Sc	3p	0.5	0.3
Ca	3p	0	+0.2 (evaluation)

Table 5.2
ΔE_{sb} shifts of Ir 4f level (eV) and I_s/I_b relative intensities [23]

Surface	E_{sb}	Experiment	Theory
Ir(100)(1 × 1)	−0.68	0.59	0.66
Ir(100)(5 × 1)	−0.49	0.98	0.94
Ir(111)	−0.50	0.84	0.80

Table 5.3
Shifts of 4f levels of RE surface atoms in metals and compounds as compared to atom levels in bulk [32], eV

RE	Metal	$REAl_2$	$REZn_2$
Eu	0.60	1.30	
Gd	0.48	0.80	
Tb	0.51	0.72	
Dy	0.53	0.76	
Tm		0.76	
Yb	0.63		0.86
	0.60 for 5p		0.80 for 5p
Lu	0.83	0.80	
	0.74 for 5p		

metals but also in GaAs and GaSb semiconductors and in diamond, where the ΔE_{sb} of the C 1s line amounts to −0.95 eV [36]. The size of the shift depends on the crystallographic plane [see, for instance, Table (5.2)].

To provide a quantitative explanation of the observed ΔE_{sb} values direct calculations of the electronic structure are presented and the thermodynamic approach is developed. Direct calculations are made for Cu(111), Cu(100), Ti(0001), and Sc(0001) (see references in [25]) and also for Tm_xSe, TmTe, YAl_2 [43]. In [43] the value ΔE_{sb} is presented as a sum of changes of the electronic contribution by ΔE_e and the change of value Δ_M in the Madelung potential for surface atoms as compared to atoms in bulk. Depending on x, ΔE_e in the Tm_xSe compound changes within the limits $+3 \div 4$ eV, and Δ_M within the limits $-(3 \div 2)x$ eV.

The fruitful nature of the thermodynamic approach to calculating ΔE_{sb} values is shown in [42]. The following expression for ΔE_{sb} has been obtained on the basis of the Haber–Born cycle within the framework of the equivalent cores method (some inconsequential components are omitted).

$$\Delta E_{sb} \simeq 0.2[E^{B}_{coh}(Z) + 1) - E^{B}_{coh}(Z)] \qquad (5.1)$$

where $E_{coh}(Z)$ is the binding energy of the element with the ordinal number Z (the indices s and b refer to surface and bulk).

This study stimulated a number of studies in the same direction. The following formula is derived in [39]:

$$\Delta E_{sb} = E_s(Z+1) - E_s(Z) \qquad (5.2)$$

where $E_s(Z)$ is the surface energy of element Z.

$$E_s(Z) = E^{b}_{coh}(Z) - E^{s}_{coh}(Z).$$

A semi-empirical formula is suggested in [40]:

$$\Delta E_{sb} = [N_{eff}(Z)/\Delta N^{b}_{eff}(Z)][\Delta H^{vap}(Z+1)^{-\Delta H^{vap}(Z)}] \qquad (5.3)$$

where N_{eff} is the effective coordination number and ΔH^{vap} is the evaporation energy.

$$\Delta N_{eff} = N^{b}_{eff}(Z) - N^{s}_{eff}(Z)$$

$$N_{eff} = N_1 + \Delta. \qquad (5.4)$$

The value of Δ depends on the type of crystal.

Equation (5.5) was proposed in [35]:

$$\Delta E_{sb} = -H_{seg}[(Z+1)\,\mathrm{in}\,Z] \qquad (5.5)$$

where H_{seg} is segregation heat.

Formulae (5.1), (5.2), (5.3), and (5.5) usually accord well with experiment (see original studies and [25]). As an illustration, in Table (5.1) experimental values for Au is compared with calculation according to (5.1), while in the other cases with calculation according to (5.3). In Tables 5.4 and 5.5 experimental values are compared with formulae (5.3) and (5.5), respectively.

The study of surface atom line shifts after adsorption is of much interest (see, for instance, [28, 34, 37].

Table 5.4
Comparison of theory and experiment for ΔE_{sb} values [40], eV

System[a]	Theory	Experiment	Theory	Experiment
Ta(III)				
ΔE^{1}_{sb}	0.35	0.40	0.53	0.63
ΔE^{2}_{sb}	0.17	0.19	0.34	0.34
$\Delta E^{1}_{sb}/\Delta E^{2}_{sb}$	2.1	2.1		
W(III)				
ΔE^{1}_{sb}	−0.39	−0.43	−0.26	−0.28
ΔE^{2}_{sb}	−0.18	−0.10	−0.06	0
$\Delta E^{1}_{sb}/\Delta E^{2}_{sb}$	2.1	4.3		

[a]Indices 1 and 2 refer to the first and second surface layers.

Table 5.5
Comparison of formula (5.5) with experiment [29], eV

System	H_{seg}	ΔE_{sb}
Ti in Sc	0.5	0.5
V in Ti	0.3	0.2
Rh in Ru	−0.2	−0.5
Pd in Rh	−0.4	−0.6

Adsorption of O_2 and H_2 on W and Ta shifts the 4f levels of surface atoms in the direction of greater binding energies, this being interpreted as transference of electron density to the adsorbate. Chemisorption of CO on Pt and W also leads to a positive shift of the levels of surface atoms (by 1.3 eV for Pt), which also can be explained by the drawing away of electronic density from the metal to CO. But in the case of the adsorption of K, which is undoubtedly a donor of electrons, there is also a 0.6 eV rise in the energy of the Pt 4f line for the surface atoms. On the basis of theoretical calculations this is explained as a result of the decrease of the number of Pt d-electrons due to the increase of s- and p-valence electrons during the formation of the Pt–K bond, although the K atom receives a positive charge in the process [34]. A theoretical explanation of shifts during adsorption can also be given on the basis of a thermodynamic approach. In particular [40], to take account of adsorption in formula (5.3) it is sufficient to add a member

$$\Delta_c^{ad} = -[E_X^{ad}(Z+1) - E_X^{ad}(Z)]\cdot\theta \quad (5.6)$$

where E_X^{ad} is the energy of X-adsorption on element Z while θ is the extent of the surface coverage. The extent of compliance of formulae (5.3) and (5.6) with experiment is shown in the right side of Table (5.4). See also [37].

The value ΔE_{sb} and also the relative intensity of the signal from surface atoms is of interest for a deep understanding of the processes of adsorption and catalysis, the electron state of surface atoms. It is also possible on the basis of these parameters to determine the mean free path of photoelectrons [17, 23, 26, 36] and to evaluate the coordinating number of atoms on the surface [23].

5.3. Study of corrosion and alloy oxidation

Reviews of the application of XPS to the study of corrosion [44, 45] and the application Auger spectroscopy to the study of electrochemical passivation [46] provide many examples of the efficiency of the methods employed (see also later references [47–53]. The combined application of several methods in surface studies is particularly effective [54].

In this section we shall dwell on the question of the oxidation of alloys. The use of XPS in this well-establised field has helped to reveal many substantial new aspects of this process, because XPS allows one to study the composition of superthin oxide films. XPS also makes it possible to distinguish between oxidation processes in the course of pulse laser treatment and lengthy oxidation [55].

The accumulated experimental material allows for certain preliminary generalizations. It is necessary to underline first of all the preliminary nature of these generalizations. The problem is that the number of studied alloys is not great, while the composition of the surface film and its in-depth changes depend on a large number of factors: alloy composition, oxidation temperature, oxygen pressure, time of oxidation, and surface roughness. One should remember that a change in surface roughness alone is sufficient for a total transformation of the results obtained. For example, in [56] it was found that the mechanical working of the surface had a substantial influence on the composition of the surface oxide films in the 5% Fe–95% Ni alloy. In a polished sample the film consists mostly (over 80%) of Ni, while in a rough sample it consists mostly (90%) of Fe_2O_3 (other conditions of oxidation being equal). This indicates that in this particular case a rough surface facilitates the migration of iron to the surface.

Note should be made of the following correlations:

(1) during the oxidation of an alloy at low temperatures the composition of the oxide film generally reflects the alloy's gross composition;

(2) even at low temperatures, and the more so at higher ones, the surface oxide film becomes enriched with the oxide of the metal whose heat of formation is higher* (refer to O_2); as a result the next layer of sample is usually enriched with the component of which the surface oxide has been depleted;

(3) with a further rise in temperature or longer heating, the topmost surface of the oxide film can become enriched with or even consist entirely of the oxide of the component diffusing to the surface (often this is the alloy's main component). It is significant that the heat of formation of free energy of this oxide on the surface can be lower than that of the oxide which initially was in the uppermost surface layer.

Here are a number of examples. Oxide films forming on stainless steel (18Cr–10Ni–1.8Mn–Fe) in the temperature range 25–800 °C, with an oxidation time ranging from 0.25 to 80 h, were studied in [57]. Oxides of iron were formed on the surface mainly at low temperatures or during brief oxidation at high temperatures. This shows that the surface film's composition is determined by the composition of the alloy itself rather than by the reactivity of the elements. High temperatures and lengthy oxidation led to an increased content of chrome and manganese in the oxide layer, agreeing with (2) above, as these metals have an increased oxygen affinity as compared to iron. X-ray photoelectron spectra made it possible also to observe the kinetics of the surface layer formation: at 500 °C the equilibrium state is achieved after 25 h of heating. This state is characterized by a greater content of chrome, and manganese, a lesser content of iron (as compared with the bulk), and an almost total absence of nickel.

During the oxidation of 13Cr–Fe steel at room temperature and with of oxygen, oxides of iron and chrome form on the surface [58], the amount of chromium oxide formed being much greater than during the oxidation of pure chrome. This points to a more extensive oxidation of chrome atoms in the alloy. The surface layer is enriched with chromium oxide. A similar phenomenon was

*For an analysis of the further presented data it shall be noted that this formation heat decreases in the series Al ~ Zr, Ti, Cr ~ Mn, Zn, Fe, Ni, Cu.

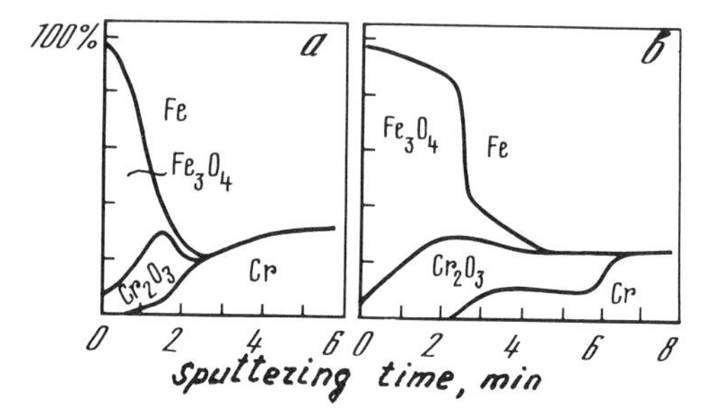

Figure 5.4. Dependence of relative phase content in surface layer of Fe–Cr oxidized alloy on sputtering time. (a) 200 °C; (b) 550 °C.

also found in [59, 60]. During the oxidation of 18Cr–10Ni–Fe steel at room temperature and with small additions of O_2 there also forms a surface layer of iron and chromium oxides enriched with chromium oxides, while atoms of nickel do not become oxidized at all [58].

During the oxidation for 1 h of 0.75Fe–0.25Cr alloy at 200 and 550 °C with $P_{O_2} = 10^{-6}$ torr, Fe_3O_4 [61] is the main surface product. The concentration of Cr_2O_3 increases with distance from the surface. The diffusion of non-oxidized atoms of iron to the surface is explained by the reactions that are taking place:

$$3Fe_3O_4 + 8Cr = 4Cr_2O_3 + 9Fe.$$

Since the number of adsorbed atoms of oxygen on the surface is greater than the number of Fe atoms at the surface, the Fe atoms on the surface become oxidized.

A study of the oxidation of the alloy 77Ni–16Cr–8Fe [56] at 100° C showed that a layer of Ni and Cr alloys forms, with the surface layer being visibly enriched with chromium oxides. During oxidation at 280° C $NiFe_2O_4$ is near the surface, while Cr_2O_3 is present at the oxide–metal border. The content of chromium oxide in the surface film increases as the temperature rises to 500° C. Chromium oxide is the surface film's main component, and with an increase of oxidation time from 0.1 to 60 min the content of Cr_2O_3 increases (at the expense of the Fe_2O_3).

In [62] the surface composition of samples of 17.4Cr–12.7Ni–2.4Mo–Fe alloy oxidized in the air was characterized by an intensity ratio Fe $2p_{3/2}$/Cr $2p_{3/2}$ of 3.4 for metals and 2.3 for oxides; in the steel 24.1Cr–24.7Ni–2.4Mo–Fe these ratios were equal to 1.6 and 1.3, respectively. The enrichment of the oxide layer with chromium oxide noticeably increases as the temperature rises above 400° C. Oxidation of steel with 6–20% Cr at room temperature and an oxygen pressure of 0.2 atm produces a film of oxides in which the chromium oxide content is in proportion to the chrome content in the alloy. The upper layer of oxides is enriched with iron while the lower layer on the metal–chrome border is presented by $FeCr_2O_4$ [63].

A layer of Al_2O_3 [64] forms on the surface during the oxidation of aluminum bronze, while Ti oxides form during the oxidation of Ti–Ni and Ti–Al alloys [65].

For completeness we shall cite some data on the oxidation of analogous alloys obtained by means of Auger spectroscopy and the diffraction of slow electrons.

It was found during the oxidation of a Fe–Cr(100) monocrystal at temperatures of 700–900 K and oxygen pressure of 10^{-9}–10^{-6} torr that there first forms a film enriched with chromium oxides, while the diffusion of iron through the chromium oxide layer begins the process of further oxidation. The phase $FeCr_2O_4$ or the solid solution $FeCr_2O_4$ and $FeFe_2O_4$ forms on the border of the metal. On the gas–oxide border the surface film consists of pure Fe_3O_4 [66]. The oxidation process of single crystal Fe–Cr(110) proceeds somewhat differently, and Fe_3O_4 does not form on the surface [66]. Oxidation of the single crystal 18.6Cr–10.6Ni–Fe at room temperature leads to the formation of Cr_2O_3, while with a rise in temperature Fe_3O_4 forms first beneath and then over the Cr_2O_3. At 500° C the Fe_3O_4 decomposes, chrome diffuses to the surface and the surface becomes enriched with Cr_2O_3. FeO(111) forms on the surface at temperatures over 750° C. Decomposition of oxides and diffusion of oxygen into the crystal bulk begins at temperatures over 900° C [67]. Similar but not identical results were obtained by the Auger method for the oxidation of the steel 9.2Ni–17.7Cr–Fe [68]. Only iron oxides are present on the surface during heating in air to 600° C, while above 700° C traces of chromium oxide appear. At 900° C, iron oxide is replaced by chromium oxide.

Using XPS, it is shown in [69] that after evaporation of aluminum onto a surface layer of oxidized chrome a reduction of Cr_2O_3 to elementary Cr takes place, owing to the oxidation of aluminum, while during the oxidation of NiCr alloy only chrome oxidizes, agreeing with the decrease of the heat of formation of oxides in the series Al, Cr, Ni.

The formation of an aluminum oxide layer is observed during the oxidation in air of aluminum alloys with 10–30 at.% Ni, while Ni atoms do not oxidize. The thickness of the Al_2O_3 layer quickly increases the temperature rises (50–450° C) or as the oxidation time increases [70]. The oxidation of permalloy (20Fe–80Ni) films [71, 72] and 76Ni–24Fe(100) single crystals [73] results in an enrichment of the surface films with Fe oxides, while the subsequent layer is impoverished of Fe atoms [71, 72]. A layer enriched with ZrO_2 forms on the surface during the first stage of oxidation of Zr_2Ni, ZrNi and $ZrNi_5$ intermetallides [74] in air at 200° C, but during additional heating nickel atoms diffuse through the ZrO_2 film and an NiO film forms on the surface. The tendency towards the formation of NiO on the surface increases with higher nickel content. Beneath the layer of oxides the alloy's composition is enriched with Ni atoms, and the degree of enrichment grows as the Zr content increases (Fig. 5.5). Analogous results were obtained during the oxidation of $ZrCo(Ni)H_{2.88}$. $HfCo(Ni)H_{2.88}$ [75, 76] and ZrHf [77, 78]. These results are close to data on the oxidation of the PbSn surface layer at room temperature and an O_2 pressure of 1 torr, depending on the duration of oxidation [53]: at first the surface is enriched with Pb, then the Pb/Sn ratio diminishes three fold in the course of 15 min (growth of the thermodynamically more advantageous SnO_2 oxide), after which an increase in the Pb/Sn ratio is observed. A small maximum is achieved after 60 min and a constant ratio after 80 min.

During the oxidation of the 59Cu–41Ni alloy at low temperatures the NiO oxide forms first in the main. With an increase of temperature to 160° C the copper(I) oxide covers the initial surface layer of NiO oxide. A further increase in

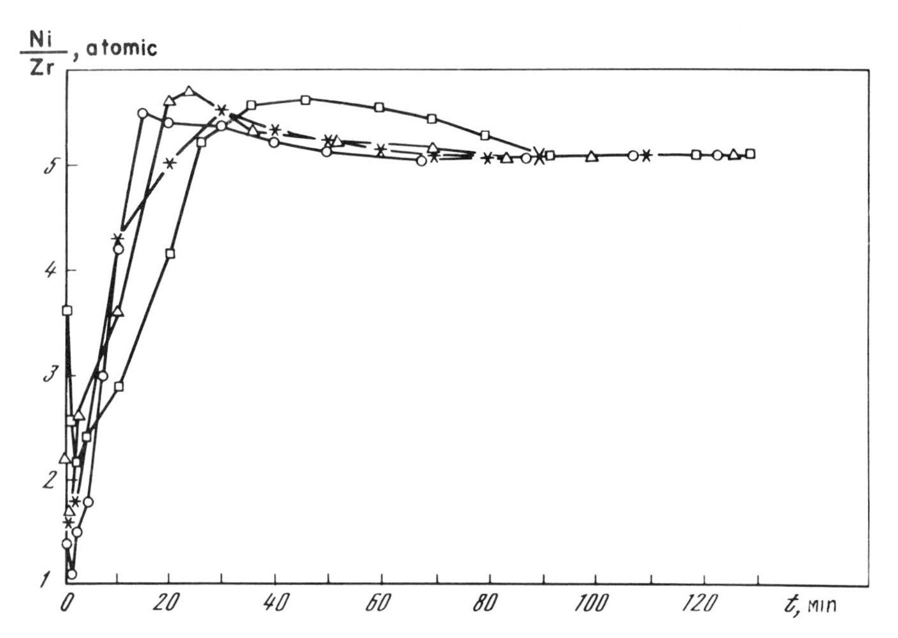

Figure 5.5. Dependence of C_{Ni}/C_{Zr} concentration ratio on $ZrNi_5$ oxidized surface on sputtering time t.
* = untreated sample; ○ = 80 min; △ = 160 min; □ = 320 min of oxidation in air at 200 °C.

temperature brings about the formation of a mixture of Cu(I) and (II) oxides in the surface layer [79].

A thin layer of Cu_2O on a thicker layer of ZnO lies on the surface of bronze 70/30 after it has been exposed to air. The thickness of the ZnO noticeably increases after heating in air, while the content of copper on the surface decreases. If the ZnO layer is dissolved, then first a layer of CuO forms on ZnO during heating, but further heating leads again to a growth of ZnO on the surface [80].

If α-bronze is oxidized at low oxygen pressures then first ZnO forms on the surface, followed by Cu_2O [81].

The patterns of alloy oxidation discussed above are not strict rules. For instance, Cu_2O oxide forms on the surface during the oxidation of the Cu–Mn alloy, even though its heat of formation is less than that of MnO. Admittedly during heating in vacuum a reaction of oxidation of Mn to MnO and reduction of Cu_2O to Cu [82] takes place on the surface.

It should be noted that knowledge of the composition of the topmost layer of oxides is extremely important for an understanding of the alloy's corrosion properties. Thus, it is shown in [83] that the noticeable increase of the corrosion resistance of the alloy Fe–Cr–Al–Y after ion implantation of Al is connected with the formation of an Al_2O_3 protective surface film. Moreover, an increase in the concentration of aluminum in the surface layer as a result of implantation facilitates the formation of this film.

Knowledge of the composition of the oxidized alloy's surface film is also important for catalysis and other processes taking place on the surface. In those instances where the surface layer accounts for a substantial part of the bulk it is necessary to take into consideration the existence of this layer. For example, specific saturation magnetization noticeably decreases when nickel particles are 10 nm in size. XPS studies have revealed that this phenomenon is connected with the NiO film on the surface of particles [84]. The patterns discovered for the oxidation of alloys also apply, at least in part, to the fluoration of alloys [85, 86].

5.4. Semiconductor surface oxidation

Intensive work in this direction has been stimulated by both its practical and its scientific importance. X-ray photoelectron spectroscopy has proved useful for solving questions concerning the composition of the oxidized layer and the first stages of oxygen adsorption in semiconductors (see, for instance, [87]). We shall limit ourselves to citing several examples.

The anode oxidation of GaAs(100) with oxygen in As_2O_3/O_2 vapours was studied in [88]. During anode oxidation the surface layer consisted of Ga_2O_3 and As_2O_3 oxides in the ratio 1:1. This ratio remained constant from the surface to the non-oxidized GaAs layer. A mixture of As_2O_3, As_2O_5 and Ga_2O_3 oxides formed on the very surface during oxidation with oxygen. The As:Ga ratio amounted roughly to 1:1. But the subsequent oxidized layer to a depth of several hundred nm consisted only of Ga_2O_3 and small quantities of GaAs and Ga_2O_3 that had failed to react (Fig. 5.6). The oxidation of GaAs in vapours of As_2O_3/O_2 resulted in the formation of $GaAsO_4$ in the surface layer followed by a mixture of As_2O_3 and Ga_2O_3 and then by the pure oxide Ga_2O_3. This data indicates that

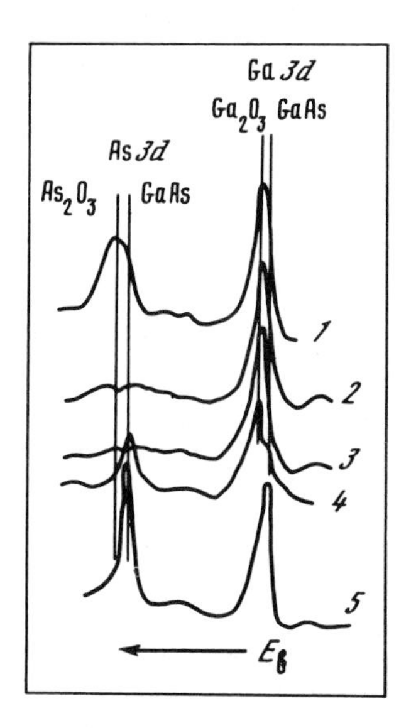

Figure 5.6. Dependence of oxidized AsGa spectra on sputtering depth.
1 = surface; 2 = 20 nm; 3 = 80 nm; 4 = 220 nm; 5 = non-oxidized GaAs.

the oxidation of GaAs by oxygen at high temperatures is accompanied by the evaporation of As_2O_3.

A detailed study of the 3.0–4.0 nm surface oxide film on GaAs with the application of chemical sputtering was made in [89]. The oxidation of GaAs took place at room temperature. Whereas the ratio of the Ga/As oxidized atoms on the surface amounted to 2:1, at the oxide/GaAs interface it was about 10:1. Also found on this interface were a thin layer of elementary arsenic and two non-equivalent sorts of Ga atom characterized by shifts of the Ga 3d line to energies 1.3 and 1.9 eV greater than in GaAs. This points to the possibility of the formation of As–O–Ga bonds (along with the formation of Ga_2O_3).

The initial stage of oxidation of GaAs(100) at exposures from 10^5 to 10^{14} L and room temperature is accompanied, according to [90], by the dissociative adsorption of oxygen and the formation of As and Ga oxides. In the process both Ga and As atoms join with oxygen already at the first stage of oxygen adsorption. In [90] the coverage amounted to about half the monolayer. In contrast to these results, in [91], with the coverage being an order of magnitude smaller, it was found that first only the oxidation of arsenic takes place, because the As 3d energy line increases while the line Ga 3d does not shift. The preference of oxygen for adsortion on atoms of arsenic and phosphorus was also noted in [92], which was devoted to a study of the oxidation of the (111) GaP and GaAs planes. But in [93] it was found that the Ga 3d line is somewhat wider in the direction of greater binding energy in GaAs(110); this can be regarded as an indication that Ga atoms take part in the interaction with oxygen. A preferential adsorption of oxygen on the Ga atom was even discovered in the case of disordered planes (110) in GaAs obtained by ion sputtering.

The following model of the initial adsorption stage was suggested in [94] on the basis of a change in the photoelectron spectrum of the GaAs(110) + O_2 system during slow and rapid heating. The oxygen atom adsorbs both on Ga atoms and on As atoms. Rapid heating leads to a desorption of O on As but does not influence the Ga–O bond (the energy and intensity of the Ga 3d line do not change, the intensity of O 1s lines and that of the shifted As 3d line decreases). During slow heating there takes place a transfer of oxygen from the As–O bond to the Ga–O bonds (an increase of the Ga 3d line shift, an intensity decrease of the shifted As 3d line and the unchanged intensity of O 1s line). It appears that for disordered planes (110) in GaAs the effect of ion sputtering is analogous to that of heating, and for this reason a predominant oxidation of Ga is also observed.

The question of the preferential adsorption of oxygen on the Ga or As atom in molecular or dissociative form was the subject of numerous studies late in the 1970s (see [90, 93–95]). For oxidation to be conducted by an ordinary non-excited O_2 molecule was a very important step in solving this question. Oxidation of GaAs(110) with a dose of up to 10^{12} L of excited oxygen leads, for instance, to As, As_2O_3, As_2O_5, and Ga_2O_3 [96]. $A_2^{III}O_3$ and $B_2^{V}O_3$ oxides form at early stages of the oxidation of InP and GaSb, using either normal or active oxygen [96]. The oxidation of A^2B^6 semiconductors is studied in [97].

The question of the molecular or dissociative mechanism of O_2 adsorption on Si during the oxidation of silicon by oxygen has also been extensively discussed

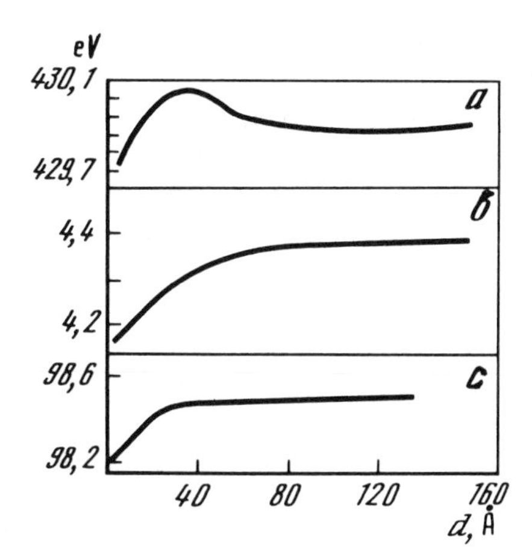

Figure 5.7. Dependence of line energies on thickness d of SiO_2 oxide on Si.
(a) O 1s – Si 2p difference; (b) Si 2p(SiO_2) – Si 2p(Si) difference; (c) energy of Si 2p(Si) line.

(see [98]). The latest detailed study [98] devoted to this question points to the dissociative nature of adsorption [98]. Of greatest interest is the question of the nature of the transition layer between Si and SiO_2 or, which is the same, the character of the thin oxide layer on silicon (see [98–101]). When the thickness of the oxide film is small the energy of O 1s and Si 2p lines in the oxide is smaller than in the massive SiO_2 (Fig. 5.7 [99]). In principle these changes can be explained by assuming the existence of a stoichiometric SiO_2 oxide if we take into account the possibility of a shift of the Fermi level in the energy gap of SiO_2 towards the valence band for the thin film SiO_2. The shift could be a result of the presence of localized states on the Si/SiO_2 interface an and increase of extra-atomic relaxation for Si and O atoms near the silicon's semiconductor substrate [98]. But usually this data is interpreted as a substantial distinction of the thin oxide film on Si from SiO_2. The following variants have been proposed depending on the thickness of the oxide film:

(1) a mixture of Si and SiO_2 at thicknesses up to 4.0 nm;
(2) continuous change of x in SiO_x near 1.0 nm;
(3) change of the oxidation state of Si from 0 to 4 in the range 0.3 to 0.7 nm;
(4) the presence of a SiO film between Si and SiO_2 (see [100, 102, 103]);
(5) the chance distribution of four-, six-, seven- and eight-member cycles of SiO_4 tetrahedrons connected by bridge acids near 3.0 nm. The length of the cycle decreases closer to the Si/SiO_2 interface. Oxides of the Si_2O_3, SiO, and Si_2O composition are near the Si/SiO_2 interface [99].

At present the question of the nature of the Si/SiO_2 interface cannot be viewed as finally solved. Within the framework of this section we shall only explain the arguments cited by the authors of [99] in support of their model, which is the most complex.

The values of Si 2p and O 1s energies depend on the angle θ O–Si between tetrahedrons. This angle is equal roughly to 120° at $n = 4$, 144° at $n = 6$, on average to more than 165° at a certain set of $n = 7$, 8, and 9, and on average is to

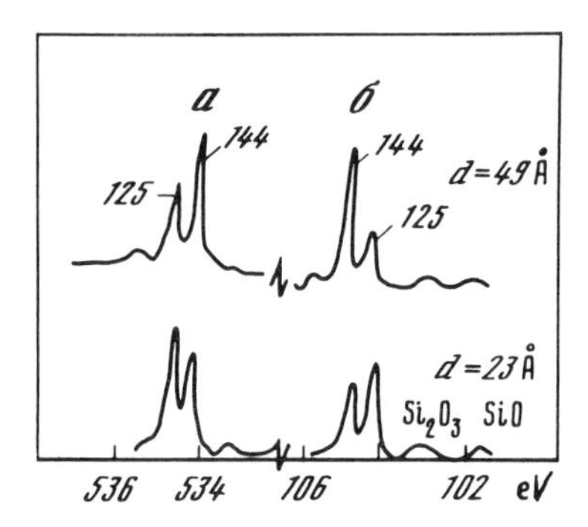

Figure 5.8. (a) O 1s and (b) Si 2p lines of SiO_2 oxide on Si for various thicknesses d of the oxide. The number near the curves are values of angle θ in degrees.

154° at $n = 5$, 7, and 8. On the whole angle θ grows as the length of the n cycle increases. With the growth of angle θ there is also an increase in the value Si 2p, while O 1s, as calculations show, decreases. After the lines Si 2p and O 1s had been divided into separate components, maxima in approximately the same energy range were found for films of various thicknesses [see Fig. (5.8), where the Si 2p and O 1s maxima connected with the substrate Si are omitted]. Each maximum was identified with its own θ angle. Data on Si 2p and O 1s intensities and energies agree with this interpretation. Along with Si 2p maxima that belong to Si(IV) atoms, the cycles also revealed maxima corresponding to lower degrees of silicon oxidation (101.8 eV for SiO and 103.1 eV for Si_2O_3).

The kinetics of surface oxidation with the forming of $SiO_x (x < 2)$ is studied in [104], while oxygen adsorption and oxidation of planes Si(111) and Si(100) are examined in [105]. The change in the length of the mean free path and/or the density of the SiO_2 layer, depending on the distance from the Si, is shown in [106].

X-ray photoelectron spectroscopy is also useful for studying other reactions on the surface of semiconductors. As an example we should note that the interaction of Al with GaAs(110) is studied in [107]. After the evaporation of a small quantity of aluminum on GaAs the Al 2p line has an energy of 73.7 eV, this being somewhat greater than in the metal: the interpretation is that there is an adsorbed Al atom on the surface. When one monolayer of Al has been deposited the energy of Al 2p increases to 74.0 eV, this corresponds to the formation of AlAs. The line of free Ga-atoms, which is substituted by Al atoms in the GaAs lattice, are registered at the same time. With a further growth of the aluminum film thickness the energy of Al 2p acquires the value of 73.1 eV, this being characteristic of the metal.

5.5. Study of fracture surfaces and alloy ageing

Auger spectroscopy is usually employed in the study of cleavage surfaces. The main advantage of Auger spectroscopy in such studies as compared to X-ray photoelectron spectroscopy is that it allows for the study of a small surface area and individual grains. The electron beam, however, can change the initial composition of the surface. For example, it is shown in [108] that initial energy scanning directly near the line Sb alone can detect the segregation of antimony on the cleavage of a steel sample. If the electron beam is on the studied point under study the antimony diffuses from this point and repeated Auger spectra do not show antimony segregation. X-ray photoelectron spectroscopy has also made it possible to record reliably the segregation of Sb (up to 10%) on the cleavage of steel containing 0.12% Sb [108]. Another advantage of X-ray photoelectron spectroscopy in such studies is that the chemical nature of compounds segregating on the surface can be investigated.

We consider as example several X-ray photoelectron studies of cleavage surface in [109–112]. A study of cleavages of the steel 37CrNi3 after various thermal treatments revealed the segregation of antimony and chrome. Additions of molybdenum drastically lessen antimony segregation on the cleavage. After various modes of tempering, segregation of elementary silicon and phosphorus was found on the cleavages of chrome steels with 5% Cr, and its content on the cleavage was two orders higher than in bulk. Chrome and manganese are on the borders of the cleavage grains in the form of carbides.

It should be noted that the chemical state of phosphorus on the surface of α-iron is studied in [113]; sulphur is studied in [114].

A study of titanium alloys of the BT9 brand with a different chemical composition made it possible to determine the chemical composition of thin films on the borders of grains with a thickness of the order of a monoatomic layer and to discover an increased segregation of boron and silicon, leading to a lessening of impact strength. Knowledge of the distribution of elements on the grain and their interaction made it possible to find methods of removing or decreasing the harmful influence of admixtures by adjusting the alloy's composition and changing it.

A study of B93 and B95 aluminum alloys revealed a segregation of nickel and silicon on the surface of the cleavage, their concentration exceeding that in the bulk 200–600-fold. The drastic worsening of the quality of semi-manufactures from these alloys in some cases is explained by the redistribution of components, in particular by the decrease of magnesium on the grain borders—this affects the composition of silicides, which in a number of instances are responsible for the destruction of products.

A study [115] of cleavages of $2\frac{1}{4}$Cr–1%Mo steel demonstrated the presence of a 0.2 monolayer of P and Fe_3C and also of Mo and Cr on the cleavage. P alone segregates on the surface of the steel during heating in the interval 500–625° C. This points to differences of segregation on the surface and borders of the cleavage.

An X-ray photoelectron study of the ageing of Al–Ag alloys was conducted in [116]. The appearance of a second Ag 3d maximum during the ageing of the Ag–Al alloy was a somewhat surprising result. The shift in relation to the metal reaches 2.3 eV. The shift for the Al 2p line was 1.5 eV. These additional maxima are connected with the emergence on the surface of the hexagonal phase Ag_2Al. Since it is known in advance that the shift of the Ag and Al inner levels in Ag_2Al should not be so large, it can be conjectured that this second surface phase is not in electrical equilibrium with the rest of the sample and the shift is partly determined by different charging of the sample's surface phase.

5.6. Study of glasses

The range of problems arising during the study of glasses and demanding for their solution the application of surface-sensitive methods such as XPS is extremely wide: chemical stability, strengthening of glass, vitrification of radioactive wastes, implantation of bioglasses, ensuring the reliable functioning of optical elements, and so forth. Some experimental results obtained in this field are studied in the present section.

First of all X-ray photoelectron spectroscopy and Auger spectroscopy were used to study the distribution of tin in the surface of float glass. Tin on both surfaces of a sheet of float glass had been discovered by the method of Auger spectroscopy by 1974 [117]. More detailed studies using XPS showed that although tin is part of the float glass surface to a depth of up to 0.5 mm, the most drastic drop in its concentration (from 30 to 2.5 weight%) occurs in a layer about 10 nm thick [118, 119]. A study of the tin state in the surface of soda calcium silicate glass has found that it either enters the glass as a modifier ion or directly infiltrates the carcass of the glass. The ratio of these two forms depends on the method of depositing the coating [120].

The distribution of glass components in the surface layers is usually determined by means of ion sputtering. Although glass is a multi-component system, according to data in [121] the graduation curves of sputtering for such components of silicate glasses as O, Ca, Mg, and Zn are of a linear character in a sufficiently wide interval of compositions, this bearing evidence to the insignificance of matrix effects. Table (5.6) [122] contains the values of effective sputtering coefficients for some components of silicate glasses (with respect to γ_{Si}). But great variability in the data is usually observed with respect to such elements as Na and K. This impedes the determination of graduation curves and thereby quantitative analysis, because of the high mobility of alkaline ions (especially Na) in glasses. For this reason the distribution of an ion in a layer altered by sputtering is strongly influenced by a large number of factors, such as sputtering speed, sample temperature, and the time span between the end of sputtering and analysis. An attempt to calculate the form of the mobile component's concentration profile and the depth of the changed layer depending on sputtering

Table 5.6
Ratio of effective sputtering coefficients of elements during Ar^+ (8 keV, 20 μA cm^{-2}) ion sputtering

$\frac{X}{Si}$	$\frac{K}{Si}$	$\frac{O}{Si}$	$\frac{Ca}{Si}$	$\frac{Zn}{Si}$
$\frac{\gamma_x}{\gamma_{Si}}$	1.96	1.01	0.64	0.42

speed, diffusion coefficient, and sample temperature was made in [122]. The author's model was based on the migration of the mobile component under the impact of the bulk charge created by ion bombardment. Calculations showed that during sputtering the mobile component's surface concentration should decrease as the temperature of the sample rises, and diminish as the speed of ion sputtering increase. The results of the calculation agree well with data on the study of potassium silicate glass (composition $20\,Na_2O$ $80\,SiO_2$, mol %). A growth in surface concentration—not only of alkaline components (sodium and potassium), but also (to a substantially lesser extent) of zinc and oxygen—was observed during the sputtering of a sample of multi-component silicate glass that was cooled by liquid nitrogen [123].

The composition of the glass surface depends very much on the method of obtaining it. In [124], when a comparison was made of the surfaces of soda silicate glass obtained by cleavage in vacuum, by firing, and by mechanical polishing, it was found that the Si content was maximum on the surface cleaved in vacuum, while O content was lowest on the cleaved surface, higher on the mechanically polished surface, and maximum on the fire-polished surface. Na content was highest on the fire-polished surface. The phenomenon of Na segregation on the surface of soda silicate glass is studied in [125]. Na surface concentration increases with heating of the sample, and reaches a constant value after 3–4 min at 300° C, after 15 min at 200° C, and after more than 30 min at 100° C. The temperature dependence of calculated Na diffusion coefficients complies with the Arrenius law but the coefficient values are somewhat smaller than those obtained by traditional isotope methods. The application of XPS has proved very effective in studying the chemical stability of glasses. An analysis of sodium concentration in-depth profiles of the surface layers of alkaline silicate glasses, after interaction with water, points to the diffusion nature of the leaching process. X-ray photoelectron studies [126] showed the disappearance of non-bridging oxygen atoms in the leached layer, complying with the formation of silanol groups Si–OH orginating in the reaction:

$$\equiv Si{-}O^-Na^+ + H_2O \rightarrow \equiv Si{-}OH + Na^+OH^-.$$

The introduction of Al_2O_3 into sodium silicate glass brings about a lessening of the share of non-bridging oxygen atoms, since in this case a part of the Na^+ ions are connected with tetrahedrons $[AlO_{4/2}]^-$ or octahedrons $[AlO_{6/2}]^{3-}$, and can depart from the glass only through ion exchange. So the chemical stability of glass should linearly increase with an increase of the content of Al_2O_3 in glass. In practice, however, chemical stability increases with the introduction of small quantities of Al_2O_3 into glass. But the increase is small. In [127], analysing glass with the composition $20Na_2O_xAl_2O_3(1-x)SiO_2$ by the XPS method, the authors arrive at the conclusion that the leaching of sodium ions connected with the polyhedrons $[AlO_{4/2}]^-$ and $[AlO_{6/2}]^{3-}$, takes place by way of ion exchange with H_3O^+ ions, whereas sodium bound with non-bridging oxygen atoms of the silicon–oxygen carcass is set free due to interaction with molecules of water. The diffusion of water is substantially slowed down in the presence of aluminum in glass.

The use of XPS has made it possible to establish also a whole range of other specificities for the interaction of glass with water. Thus, the dependence of the corrosion speed of alkaline silicate glass on the ratio of the total glass surface area to the volume of the solution in which it is contained has been established [128]. A growth in this ratio leads to a drastic increase in the corrosion speed of glass. So both an increase of surface area as a result of poor surface treatment and a decrease of the volume of the corrosion solution contained between glass articles that are in direct contact, for instance between two sheets of glass, equally result in a higher corrosion speed in the glass.

The process of ion-exchange strengthening of sodium silicate glass in KNO_3 melt was studied in [129, 113]. The discovered 'blocking' effect of the K^+_{glass}–Na^+_{melt} ion exchange is explained by the formation on the glass surface of a layer enriched with calcium sorbing from the KNO_3 melt and containing calcium ions as an admixture. The presence of calcium is found even on the surface of glass samples treated in melts of highly purified potassium and sodium nitrate (content of admixture cations less than 0.001 wt.%). An additional introduction of Mg or Zn nitrates into the calcium-'polluted' KNO_3 melt lessens the 'blocking' effect. In the process Mg^{2+} and Zn^{2+} ions enter the glass to a substantially greater depth than Ca^{2+} ions.

Besides determining composition, XPS makes it possible to study the specificities of the electron structure of the glass surface layers. Thus, in [130] two components corresponding to the 'normal' Si-atoms in glass and Si-atoms in the group $Si(CH_3)_2$, with the share of the latter increasing as the angle α between the sample plane and the direction of photoelectron registration decreases, are distinguished in the Si 2p line after dimethyldichlorsilan $[Cl_2Si(CH_3)_2]$ treatment of the silicate glass surface.

In a number of instances an analysis of the O 1s photoelectron line makes it possible to determine the short order structure in glasses of various composition. The change in the ratio of non-bridging to bridging oxygen atoms after the introduction of some component into silicate glass makes it possible to determine whether the introduced component is a modifier or a grid former. For example, for sodium silicate glass with a composition $x\mathrm{Na_2O}(1-x)\mathrm{SiO_2}$, this ratio should be $(2-3x)/2x$, this agreeing with XPS data. By analogy, Li, K, and Cs alkaline ions are modifiers in silicate glasses, with the maximum chemical shift between bridging and non-bridging oxygen atoms being observed for caesium and the minimum shift for lithium glasses [131]. At the same time, in mixed alkaline silicate glasses (for instance, in sodium caesium silicate glass) the O 1s spectra cannot be regarded as a simple superposition of the spectra of individual components, that is the non-bridging oxygen atoms cannot be unequivocally ascribed to some definite sort of modifier ions, and the formation of structural units of the type

$$\equiv\mathrm{Si{-}O^-}\begin{matrix}\mathrm{Na^+}\\ \mathrm{Cs^+}\end{matrix}\mathrm{O^-{-}Si}\equiv$$

is most probable (in the case of sodium caesium silicate glass).

As noted above, the introduction of Al_2O_3 into sodium silicate glass results in a lessening of the share of non-bridging oxygen atoms. The total disappearance of the corresponding signal [131] is observed during the ratio of molar shares $Al_2O_3/Na_2O = 0.7$, testifying to the formation of octahedrons $[AlO_{6/2}]^{3-}$, that form a large number of bridge bonds, along with tetrahedrons $[AlO_{4/2}]^{-}$. The same patterns were observed in sodium gallium silicate glasses. The form of the O 1s spectrum for sodium indium silicate glass does not allow for clearly establishing the state of indium in this class—the component with a lesser E_b does not disappear as in the case of Al_2O_3 and Ga_2O_3, while the shift between non-bridging and bridging oxygen atoms is substantially smaller than in the case of sodium silicate glass. But after interaction with water (over several minutes at $T = 20°$ C) the proportion of non-bridging oxygen atoms lessens, as a result of leaching, by a value that corresponds to the number of sodium ions in the glass. This shows that all sodium ions are bound to the non-bridging oxygen atoms of the silicon–oxygen carcass and not to the octahedrons $[InO_{4/2}]^{3-}$, that is the In^{3+} ions in this case are modifiers and not grid formers [133].

Interesting results were obtained in [132] during the study of phosphate glasses with composition $xR_2O(1-x)P_2O_5$, where R = Li, Na, K, Cs. The O 1s photoelectron line resolves into three components corresponding to bridging and non-bridging oxygen, and oxygen atoms attached to phosphorus by a double bond. An analysis of O 1s spectra reveals the same patterns as for alkaline silicate glasses. A growth of alkaline cation electrostatic field in the series Cs to Li leads to a decrease of the chemical shift in O 1s, i.e. the O 1s values for the various oxygen atoms types become similar.

Non-bridging oxygen atoms are absent in alkaline germanate glass with an alkaline oxide content reaches 20 mol%, i.e. the structure is formed from $GeO_{4/2}$ octahedrons. When the alkaline oxide content reaches 33%, germanate and silicate glass become isostructural. This is confirmed by the respective X-ray photoelectron spectra [134].

Detailed information on the application of surface-sensitive methods to the study of glass can be found in reviews [135, 136].

X-ray photoelectron spectroscopy is widely used, not only in the study of silicate but also in studying metallic glasses (amorphous alloys). See for instance [137, 138], and references in these papers.

5.7. Analysis of the surface of minerals: flotation

The surface composition of minerals may differ from bulk composition for a number of reasons. XPS is a convenient method of studying the mineral surfaces because it can help determine the quantitative correlation of elements on the surface and the character of the chemical compound. The surface composition of minerals is especially important to an understanding of their flotation potential, and XPS has recently begun to be applied to the determination of surface compounds formed on minerals in the course of flotation (see [139, 140] and references cited therein).

Let us consider three examples: an analysis of surface compounds formed on ZnS as it is treated with $CuSO_4$ [139]; the analysis PbS treated with $K_2Cr_2O_7$ [139]; and the composition of native gold.

5.7.1. *Activation of sphalerite ZnS with $CuSO_4$*

Results of the analyses of pure ZnS and of ZnS with appreciable traces of Fe(ZnS_{Fe}) are presented in Table (5.7). Data for compounds used to identify surface compounds are given in Table (5.8). The data given in the two tables prompt the following conclusions. Cu_2S is formed on the surface of ZnS treated with $CuSO_4$ in H_2O. This is shown by the energy values of the Cu $2p_{3/2}$ line. In addition, the Cu 2p line in the compound of the mineral surface does not contain satellites, which usually are observed for the Cu 2p line in CuS. The surface layer therefore contains only Cu_2S and no CuS. In addition to Cu_2S, SO_4^{2-} ions due to

Table 5.7
Binding energies (eV) and atomic ratios for ZnS treated with $CuSO_4$

$CuSO_4$ concentration, $mg\,ml^{-1}$	Zn $2p_{3/2}$	S 2p	Cu $2p_{3/2}$	S/SO_4	Zn/Cu
		ZnS			
—	1022.0	162.1	—	—	—
20	1022.1	162.2/169.5[a]	932.4	5.5	4.6
50	1022.1	162.2/169.5	—	5.5	4.6
100	1022.1	162.2/169.5	932.4	5.2	4.0
200	1022.1	162.2/169.5	932.4	5.2	4.1
500	1022.1	162.2/169.5	932.4	4.7	3.5
		ZnS_{Fe}			
—	1022.0	162.2	—	—	—
20	1022.0	162.2/169.5[a]	932.3	19.1	3.5
50	1022.0	162.2/169.5	932.3	16.8	3.3
200	1022.0	162.2/169.5	932.3	15.8	2.9
500	1022.0	162.2/169.5	932.3	15.2	2.8

[a]Two S 2p peaks are observed.

Table 5.8
Binding energies (eV)

Compound	Binding energies			
	Spectral line	Energy, eV	Spectral line	Energy, eV
Cu_2S	Cu $2p_{3/2}$	932.6	S 2p	162.6
CuS	Cu $2p_{3/2}$	935.2	S 2p	162.2
$CuSO_4 \cdot nH_2O$	Cu $2p_{3/2}$	935.8	S 2p	169.5
PbO	Pb $4f_{7/2}$	138.3		
$PbSO_4$	Pb $4f_{7/2}$	139.6	S 2p	168.8
$Pb(OH)_2$	Pb $4f_{7/2}$	138.6		
$K_2Cr_2O_7$	Cr $2p_{3/2}$	580.1	K 2p	293.0
Cr_2O_3[a]	Cr $2p_{3/2}$	577.7		

[a]For $Cr(OH)_3$ or CrO(OH); on the surface the value in Cr_2O_3 equals 576.1 eV.

oxidation were discovered on the ZnS surface, as shown by the energy of S 2p line equalling 169.5 eV. The SO_4^{2-} anion is not connected with the Cu^{2+} cation because in that case we should expect the Cu 2p line to have an energy of 935.8 eV, the intensity of which must be comparable to that of the line produced by Cu_2S with an energy of about 932.5 eV. The SO_4^{2-} anion seems to be connected with zinc.

As the concentration of $CuSO_4$ rises the Cu_2S and SO_4^{2-} content of the ZnS surface increases (Table 5.7). The iron content of ZnS substantially influences the content of Cu_2S and SO_4^{2-} on the surface: the larger the Fe content of ZnS the larger the surface Cu_2S content and the smaller the surface SO_4^{2-} content.

5.7.1. *Depression of galena (PbS) with $K_2Cr_2O_7$*

Data in Tables (5.8) and (5.9) prompt the conclusion taht $Pb(OH)_2 + Cr_2O_3$ or $PbCr_2O_4$ are formed on the surface of PbS. It is difficult to decide with any degree of certainty on the character of the compound on the basis of XPS data because the Pb $4f_{7/2}$ and Cr $2p_{3/2}$ lines should have close values in both cases. It should be stressed that, according to XPS data, the formation of $PbCrO_4$ on the surface of PbS is ruled out, although the formation of chromate is presumed fairly often. As the concentration of $K_2Cr_2O_7$ grows, the Pb/Cr ratio approaches unity (Table 5.3). The SO_4^{2-} ion was also found on the PbS surface, at a concentration of about 20% of that of S. The appearance of the SO_4^{2-} is connected with the oxidation of lead sulphide in aqueous medium on account of dissolved oxygen.

5.7.2. *Native gold surface [141–145]*

Native lode and placer gold contains up to 50% silver. Since there is no metal smelting in the course of gold formation, silver can be distributed unevenly in native gold; the surface of gold particles, in particular, may be poor or rich in silver. There are found in relevant publications indirect indications of both very high and extremely low silver content on the surface of native gold.

Studies [141, 142] established by the XPS method the C^s(Ag/Au) ratio on the surface for gold particles from different deposits. Special emphasis was laid on the selection of particles from rock in a way precluding any change in the surface composition. In particular, only particles extracted from rock by hand and not at

Table 5.9
Binding energies (eV) of PbS atoms after treatment with potassium bichromate

$K_2Cr_2O_7$ concentration, mg l^{-1}	Cr $2p_{3/2}$	S 2p	Pb $4f_{7/2}$	Atomic ratio Pb/Cr
—	—	160.9	137.7	—
20	577.4	161.4	138.5	2.0
50	577.4	161.4/168.0[a]	138.5	1.3
100	577.4	161.4/168.0	138.5	1.1
200	577.4	161.4/168.0	138.5	1.1
500	577.4	161.4/168.0	138.5	1.1

[a]Two S 2p maxima were recorded.

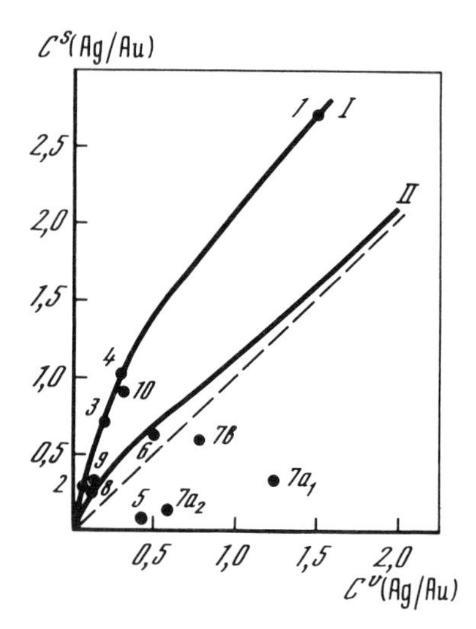

Figure 5.9. Relationship between the surface C^s(Ag/Au) and volume C^v(Ag/Au) ratios of Ag and Au concentrations. Conventional curves for lode gold (I) and alloys (II).

all subjected to chemical agents were analysed. The results of the analysis are presented in Fig. (5.9): 7b is native lode gold and $7a_1$ and $7a_2$ are placer from the same deposit. Particle size: $0.4 + 0.31(7a_1)$, $-0.1 + 0.08(7a_2)$. Data in Fig. (5.9) prompt the following conclusions:

1. The surface of lode gold (deposits 1–4) is enriched with silver and this enrichment is far higher than it is in Ag–Au alloys with the same composition, and also is found in the case of large Ag concentrations. The relevant points are connected by a smooth curve is given for the sake of illustration but the available experimental data are insufficient to claim the existence of any general dependence of surface concentration C^s(Ag/Au) on volume concentrations C^v(Ag/Au) for particles of lode gold regardless of the character of the deposit in question.

For instance, metallic gold particles from a deposit with very unusual geological conditions had a remarkably high enrichment of the surface with silver: the C^s(Ag/Au)/C^v(Ag/Au) ratio was about 10, far higher than that for particles from deposit 1 with roughly similar C^v(Ag/Au) values.

A study of samples of lode gold by means of ion sputtering established that the thickness of the Ag-enriched layer is about 5–10 nm. Simulation experiments demonstrated that this layer is formed through the diffusion of silver towards the surface and its fixation on the surface of silver particles in the form of Ag_2O.

2. The surface of placer gold (deposits 5, 6, and 7a) is slightly poor in Ag.

It is noteworthy that the surface silver content for particles from deposit 6, which were exposed to far smaller mechanical effects in the process of transfer (buried placer), was far higher than that for particles from deposit 5 (open placer). This can be viewed as an indication of the mechanical removal of silver from the surface of particles in the process of metal migration in placers. Simulation experiments confirm this conclusion.

Studies of samples of placer gold by means of ion sputtering have demonstrated that in some samples the surface layer, with a thickness of several micrometres, does not contain any silver. In the general case, the gold-enriched layer also is several micrometers thick.

3. Particles of lode gold taken from oxidation zones (deposits 7b, 8) have displayed an Ag impoverishment of the Au surface as compared with lode gold of the same composition not extracted from oxidation zones. However, no relative Ag impoverishing was discovered in a sample from the oxidation zone of deposit 10.

A sample of secondary gold from deposit 9 also was analysed and its surface proved to be enriched with silver.

Available experimental data therefore indicate that the value $C^s(Ag/Au)$ for native gold particles depends substantially on the deposit type (lode or placer) and that, moreover, it is very important in the case of lode gold whether it is extracted from an oxidation or non-oxidation zone.

5.8. XPS analysis of lunar regolith

Studies [146, 147] discovered that the surface of lunar regolith particles produces unusual Fe 2p spectra (Fig. 5.10), which peak out at around 707 eV, where the maximum of metallic iron is expected. Since regolith had been kept in the earth atmosphere for a long time,* it was to be expected that metallic iron had been oxidized and that this maximum should be absent. An Fe 2p maximum in FeS_2

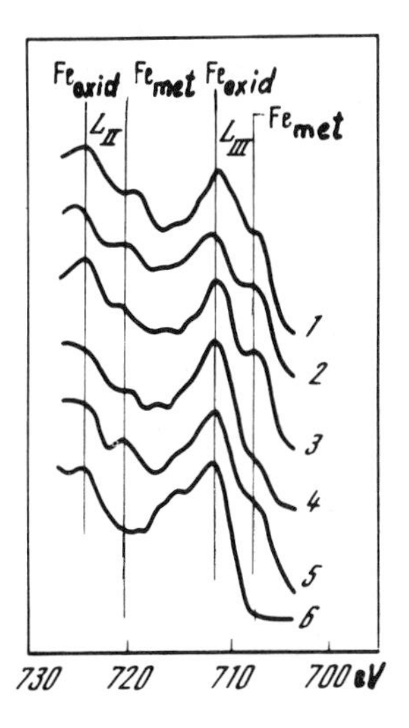

Figure 5.10. Fe 2p line. Numbers denote sources of different regolith samples: (1–3) Luna-20; (4) Apollo-12; (5) Luna-16; (6) Luna-16 (after grinding).

*Subsequent studies demonstrated that the spectrum is little influenced even if the sample is exposed to atmospheric conditions over many years.

can occur in the same spectral zone but a comparison of the Fe 2p and S 2p signal intensities demonstrated that this maximum cannot be explained by the presence of FeS_2 on the surface. Later the occurrence of non-oxidized iron on the surface of lunar regolith particles was confirmed in studies [148, 149] while studies [150–152] also discovered non-oxidized silicon and titanum forms. Non-oxidized forms were found in virtually all the samples of lunar regolith from different parts of the Moon [146–153].

Non-oxidized forms are finely disperesed in the surface layer of particles. It suffices to grind a sample to have the signals of non-oxidized forms disappear (see Fig. 5.10). Ion sputtering [154, 155] demonstrated that non-oxidized forms are found in a surface layer of a thickness of up to 100–200 nm. The share of non-oxidized Fe, Ti, and Si amounts to some 10.5 and 5% of oxidized forms on the surface and gradually diminishes into the bulk of the particle. The layer containing non-oxidized iron is thicker than it is for other non-oxidized elements. Another specific characteristic of lunar regolith is a marked difference between the surface and volume compositions, first discovered in [156] and confirmed in [148–150]. Table (5.10) gives data for different samples of lunar regolith: the surface composition of lunar regolith particles is characterized by a higher content of SiO_2 and lower content of CaO (and MgO [148, 149]). It is noteworthy that the results of [146, 156] are quantitatively adequate [152]. Changes in the composition only affect the outer layer of particles and even grinding causes the effect under review to disappear.

It can be considered established with certainty today that the unusual surface characteristics of lunar regolith are explained by the effect of 'solar wind', i.e. a flux of ions (predominantly protons) and micrometeorite bombardment; how-

Table 5.10
Surface and volume concentrations of components in regolith

	Concentration, %			
Sample[a]	SiO_2	FeO	Al_2O_3	CaO
Luna-16				
Surface	52	18	12	8
Volume	41.7	16.8	15.3	12.2
Luna-20				
Surface	52	7	23	8
Volume	42.3	7	22.7	17.1
Luna-24				
Surface	54	17	13	6
Volume	45	17.8	13.8	10.7
Apollo-11				
Surface	50	17	14	9
Volume	43	16	13	12
Apollo-12				
Surface	54	15	12	9
Volume	42	17	14	10

[a]Sample numbers are given in [152].

ever, there is no clear explanation of the abnormally low oxidation speed, for example of iron, under the earth's atmospheric conditions. Hypotheses under debate include the presence of superfine films of oxides, carbides or silicides [152], the presence of an energy gap in the electron structure of small clusters of Fe atoms which substantially influences the metal–oxygen interaction [152], a high degree of purity of iron (or an unusual composition of inhibiting admixtures) [146, 147], etc. Another possibility is the specific effect of the exposure of the surface to the flux of 'solar wind' particles, resulting in the formation of a protective coating, amorphization, compression, suppression of active centres, etc. Experiments [157, 158] indeed indicate an increase in resistance to acid-induced corrosion and surface oxidation in metallic specimens exposed to special ion treatment (see also review [159]). The mechanism of the protective effect of ion treatment is not quite clear. It should be borne in mind, however, that the conditions of exposure to the 'solar wind' and ion bombardment are appreciably different. The protective effect of ion bombardment can be caused e.g., by the formation of a protective coating from a layer of impurities, which are not found under conditions in space.

Abnormal surface properties are likely to be found in other minerals found in space and not only in lunar regolith (see for example [160], dealing with Martian regolith).

5.9. Study of adhesion

XPS offers insights into the nature of polymer–metal adhesion forces, into the character of the surface following the rupture of polymer and of metal inside polymer, or on the polymer–metal boundary, etc. (see for example review [161]). There follow several examples.

Papers [162, 163] analyse the copper–polymer rupture surface. The polymer surface was pre-treated, including the deposition of a solution of bivalent tin chloride, prior to galvanic copper deposition. The intensity of the Sn 3d signal on rupture surfaces along with other spectral data was used to determine the character of the rupture surface. The intensity of the Sn 3d signal on both rupture surfaces was infinitesimally small for the copper–polythene–tetraphtholate system; moreover, the ratio of intensities Cu 2p/Cu 3p indicated the presence of a thin polymer film on metal. The rupture had therefore occurred inside polymer. The rupture in the copper–polyurethane system, conversely, occurred along the copper–polymer boundary and the transition layer remained on the copper side of the rupture because the composition of the polymer surface corresponded to untreated polymer while the copper rupture surface contained a vast quantity of tin. Reference [164] also illustrates with the help of X-ray photoelectron spectra that the rupture in two investigated polymers on aluminum occurs inside polymer. Reference [165] demonstrates the role of oxygen in the formation of a metal–polymer bond: the presence of oxygen improves adhesion.

Study [166] looks into the causes of an increase in the adhesion of the oxidized surface of a polyvinyl alcohol specimen to metals. To this end copper was being gradually deposited onto the surface. With very fine layers (0.005 of monolayer), the Cu $2p_{3/2}$ energy was 1.2 eV higher than it is in metallic Cu (due to smaller extra-atomic relaxation). The Cu 2p value in a two atomic layer equalled that of metal. As copper was added, the C 1s and O 1s intensities were changed and the intensity of the C 1s level with an energy of 286.6 eV was changing faster than in the case of an energy of 285 eV. A new O 1s peak with an energy of 530.6 eV appeared. These changes are interpreted as a result of the formation of a surface complex incorporating copper, which increases adhesion.

Reference [80] studies compounds on the rubber–bronze boundary. It is a practical problem because rubber used in brakes cannot be allowed to come into contact with steel. That is why rubber is coated with a layer of electrolytic bronze. An analysis of the composition demonstrated that ZnS and Cu_xS sulphides are formed on the vulcanized rubber–bronze boundary and that the presence of Cu_xS contributes to good adhesion between rubber and bronze. (There is no adhesion between metallic zinc and rubber.) The amount of Cu_xS in the transition layer should be controlled, however, because a surplus makes the rubber–bronze transition layer brittle.

An analysis of a similar problem in [167] showed the undesirability of overheating in the production of brass-coated steel cords as it causes an increase in β-brass content on the surface with consequent smaller adhesion to rubber.

5.10. Study of radiation damage

The XPS method makes it possible to determine changes in the quantitative composition and to study the progress of chemical processes on the surface under the influence of radiation of different types, such as gamma and X-ray radiation and ion and electrom beams (see review [168]; the effect of ion beams is considered in Chapter 3).

We will look at a few examples of research into changes in the surface composition under the effect of radiation. Reference [169] traces changes in $NaNO_3$ subject to X-ray radiation. Peaks related to NO_2^-, NO^-, and N^- and N^{3-} were discovered in the N 1s spectrum in addition to the peak corresponding to the NO_3^- group. The composition of the surface film depends on temperature: emissions of N_2, O_2, and NO were detected as the temperature rose from that of liquid nitrogen to ambient temperature. Reference [170] studied the kinetics of the reduction of Cu(II)–Cu(I) salts under the influence of X-ray radiation.

An XPS analysis of changes in the surface composition under the effect of electron beams is of special interest because in this case radiation damage often affects only a very thin surface layer, and the effects can only be recorded by the XPS method. A series of studies by Sasaki [171–173], for instance, consider the disintegration of lithium nitrate, sulphate, chromate, tungstate, chlorate, bromate, and periodate. Li_2SO_4 was irradiated with electrons having energies of 03–1.6 keV. As the radiation dose I_c grew the S 2p and O 1s spectra changed (Fig. 5.11 [171]). The presence of Li_2SO_3, Li_2S, S, Li_2O, and adsorbed O_2^+, O_2^- and O_3^- is demonstrated on the basis of the energies of individual components. Relative intensities make it possible to determine the kinetics of the formation of these products.

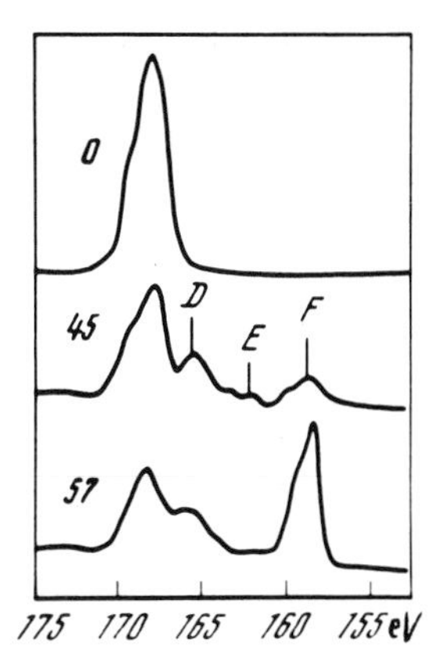

Figure 5.11. Dependence of variations of the S 2p line in Li_2SO_4 on the radiation dose I_c, in curie (figures next to curves).
Peaks: D = SO_3^{2-}; E = S; F = S^{2-}.

An interesting phenomenon was discovered in the electron irradiation of Li_2WO_4: the entire spectrum was shifted by 2 eV towards smaller energies. The shift kept decreasing over 20 h until none of the lines returned to the original state. The phenomenon is explained by the negative charging of the surface as a result of irradiation.

Reference [174] states on the basis of X-ray photoelectron spectra that the impact of electrons (2 keV) on bromosodalite leads to a drastic decrease in the content of sodium and bromine on the surface. Heating, however, helps to increase the concentration of sodium and (partially) bromine on the surface.

It should be added in conclusion of this section that the X-ray radiation used to obtain X-ray photoelectron spectra can in itself damage the specimen, specifically the AlK_α or MgK_α lines rather than the bremsstrahlung [175].

5.11. Study of diffusion

A quantitative analysis of surface composition allows the use of X-ray photoelectron and photoelectron spectra to study diffusion on the surface. The simplest variant is to study the diffusion of a surface film into the substratum (see [176] and the literature quoted there). As an illustration let us study the results of [176], that investigated the diffusion of Pd deposited on the (0001) facet of a Zn crystal.

The determination of the diffusion coefficient D on the basis of XPS data is founded on the following ratios. The ratio $R(t) = I(t)/I(0)$ is measured, where $I(t)$ is the intensity of signal Pd 4d at time t. If the thickness of the Pd layer on Zn is less than the escape depth L of the photoelectron, then the initial Pd distribution $n_{Pd}(x, t, D)$ can be regarded as a δ function. In the general case for the function $R(t, D, L)$ we have

$$R(t, D, L) = \int_0^\infty n_{Pd}(x, t, D) e^{-x/L} dx \qquad (5.7)$$

where x = normal coordinate to the surface while n_{Pd} should be obtained as a result of solving the diffusion equation. The δ-function can be taken as n_{Pd} for a coating with less than three monolayers. Using the ratio (5.7) it is possible to calculate $R(t, D, L)$ as a function of $\sqrt{Dt}$ and to determine $\sqrt{Dt}$ and thereby D, on the basis of experimental values $R(t)$. In this example experiment demonstrated that the value D changes with time: the diffusion of Pd into Zn at first proceeds faster. An evaluation of the diffusion coefficient of Zn into Pd was also made. It was taken into account that this diffusion proceeds much faster than diffusion of Pd into Zn and for this reason the growth of the Zn 3d line intensity is connected only with Zn diffusion. In this case the stream $\mathcal{T}_{Zn}$ of Zn atoms on the surface per unit of time equals

$$\mathcal{T}_{Zn} = \frac{N_{Zn}}{\Delta t} = D\frac{N_{Zn}(S) - N_{Zn}(O)}{S} = \frac{DN_{Zn}(S)}{S} \tag{5.8}$$

where N_{Zn} is the number of Zn atoms per unit of surface that reached the surface during time Δt, $N_{Zn}(S)$ is the Zn concentration on the interface Pd/Zn, while S is the average thickness of Pd film at time interval Δt. It is easy to determine the value N_{Zn} by the growth of the Zn 3d line intensity over the period Δt. $N_{Zn}(S)$ is roughly equal to the Zn concentration in the metal bulk. This makes it possible to determine D.

References

1. Haneman, D. (1968) Electron paramagnetic resonance from clean single-crystal cleavage surface of silicon. *Phys. Rev.* **170**, 705–718.
2. Westphal, D. and Goldman, A. (1980) Polarization dependent photoemission from d-line surface states on Cu. *Surf. Sci.* **95**, L249–L253.
3. Pessa, M., Asonen, H., Rao, R. S. *et al.* (1982) Surface electronic states in alloys. *Surf. Sci.* **117**, 371–375.
4. Himpsel, F. J. and Eastman, D. E. (1979) Photoemission studies of intrinsic surface states in Si (100). *J. Vac. Sci. Technol.* **16**, 1297–1299.
5. Ludeke, R. and Ley, L. (1979). Ga As surface states observed by X-ray photoemission. *J. Vac. Sci. Technol.* **16**, 1300–1301.
6. Citrin, P. H., Wertheim, G. K. and Baer, Y. (1979) Core level binding energy and density of states from the first atomic layer of gold. *J. Vac. Sci. Technol.* **16**, 653 (Abstr.).
7. Fukuda Y., Honda, F., and Rabalais, J. W. (1980) XPS and UPS study of the valence band structure and chemisorption Ti(0001). *Surf. Sci.* **91**, 165–179.
8. Heinemann, P., Hermanson, J., Miosga, H., Neddermeyer, H. (1979) Photoemission observation of new surface state band on Cu (110). *Surf. Sci.* **85**, 263–268.
9. Rowe, J. E., Traum, M. M., and Smith, N. V. (1974) Measurement of the angle of dangling-bond photoemission from cleaved silicon. *Phys. Rev. Lett.* **33**, 1333–1335.
10. Colbert, J. and Shevchik, N. J. (1979). Bulk interpretation of angle resolved photoemission spectra from GaAs(110). *J. Vav. Sci. Technol.* **16**, 1302–1306.
11. Jakobi, K., Muschwitz, C., and Ranke, W. (1979) Angular-resolved UPS of surface states on GaAs(111) prepared by molecular beam epitaxy. *Surf. Sci.* **82**, 270–282.
12. Knapp, J. A. and Lapeyere, G. J. (1976) Angle-resolved photoemission studies of surface states on (110) GaAs. *J. Vac. Sci. Technol.* **13**, 757–760.

13. Huijser, A., Van Laar, J., and Van Roog, T. L. (1978) Angle- resolved photoemission from GaAs(110) surface states. *Phys. Lett.* **65A**, 337–342.
14. Colbert, J. and Shevchik, N. J. (1979) Bulk interpretation of the angle-resolved photoemission spectra GaAs(110). *J. Vac. Sci. Technol.* **16**, 1302–1305.
15. Chadi, D. J. (1978) Surface structure and orbital symmetries of (110) surface states of GaAs. *J. Vac. Sci. Technol.* **15**, 631–636.
16. Guichar, G. M., Sebenne, C. A., and Thuault, C. D. (1979) Electronic surface states on cleaved GaP(110): initial steps of the oxygen chemisorption. *J. Vac. Sci. Technol.* **16**, 1212–1215.
17. Leonhardt, G., Neumann, H., Kosakov, A. *et al.* (1977) Investigation of the electronic structure of V_2VI_3 compounds by means of X-ray emission and photoelectron spectroscopy. *Phys. Scr.* **16**, 448–451.
18. Nishitani, N., Aono, M., Tanaka, T. *et al.* (1980) Surface states on the LaB_6 (100), (110), (111) clean surface studies by angle-resolved UPS. *Surf. Sci.* **95**, 341–358.
19. Uhrberg, R. I. G., Hansson, G. V., Nicholls, J. M. *et al.* (1982) Experimental surface state bands of the Si(111)2 × 1 reconstructed. *Surf. Sci.* **117**, 394–404.
20. Hansson, G. V., Uhrberg, R. I. G., and Nicholls, J. M. (1983) Photoemission studies of surface states on Si(111)2 × 1. *Surf. Sci.* **132**, 31–39.
21. Straub, D., Dose, V., and Altmann, W. (1983) Investigation of intrinsic unoccupied surface states of CuP(110) by inverse photoemission. *Surf. Sci.* **133**, 9–14.
22. Citrin, P. H., Wertheim, G. K., and Baer, Y. (1978) Core-level binding energy and density of states from the surface atoms of gold. *Phys. Rev. Lett.* **41**, 1425–1428.
23. Van der Veen, J. F., Himpsel, F. Y., and Eastman, D. F. (1980) Structure dependent 4f-core-level binding energies for surface atoms on Ir(111), Zr(100)–(5 × 1) and metastable Ir(100)–(1 × 1). *Phys. Rev. Lett.* **44**, 189–192, 553.
24. Desjonqueres, M. C., Spanjaard, D., Lassailly, V., and Guillot, C. (1980) On the origin of the variation of the binding energy shifts of core levels between surface and bulk atoms in transition metals. *Solid State Commun.* **34**, 807–810.
25. Citrin, P. H. and Wertheim, G. K. (1983) Photoemission from surface-atom core levels, surface densities of states and metal-atom clusters: a unified picture. *Phys. Rev.* **B27**, 3176–3200.
26. Citrin, P. H., Wertheim, G. K., and Bayer, Y. (1983) Surface-atom X-ray photoemission from clean metals: Cu, Ag, Au. *Phys. Rev.* **B27**, 3160—3175.
27. Hecht, M. H., Viescas, A. Y., Lindau, I. *et al.* (1984) Intrinsic surface binding energy shifts in Yb metal. *J. Electron Spectrosc. Ref. Phen.* **34**, 343–353.
28. Apai, G., Baetzold, R. C., Shustorovich, E., and Jaeger, R. (1982). Surface core level shifts for clean Pt(III) and Pt(III) with chemisorbed CO. *Surf. Sci.* **116**, L191–L194.
29. Neddermeyer, H., Misse, U., and Rupieper, P. (1982) Photoemission study of the surface electronic structure of Si(111)1 × 1 and Si(111)7 × 7. *Surf. Sci.* **117**, 405–416.
30. Gerken, F., Barth, J., Kammerer, R. *et al.* (1982) Surface shifts of rare earth metals. *Surf. Sci.* **117**, 468–474.
31. Johansson, L. I., Flodstrom, A., Hornstrom, S. E. *et al.* (1982) Surface shifted core levels used as a tool to identify surface segregation in Eu–Au and Yb–Au alloys. *Surf. Sci.* **117**, 475–481.
32. Kaindl, G., Schneider, W. D., Laubschat, C. *et al.* (1983) Core-level shifts and valence change at the surface of intermetallic rare-earth systems. *Surf. Sci.* **126**, 105–111.
33. Erbudak, M., Kalt, P., Schlapbach, L., and Bennemann, K. (1983) Surface core-level shifts at the end and beginning of the transition metal series. *Surf. Sci.* **126**, 101–104.
34. Apai, G., Baetzold, R. C., Yupiter, A. Y. *et al.* (1983) Influence of acceptor and donor adsorbates on Pt surface core-level shift. *Surf. Sci.* **134**, 122.

35. Feibelman, P. J. (1983) Comparison of calculated surface core-level energy shifts with empirical heats of segregation. *J. Vac. Sci. Technol.* **1**, 1101–1102.
36. Pate, B. B., Oshima, M., Silbermann, Y. A. *et al.* (1984) Carbon 1s studies of diamond (111). *J. Vac. Sci. Technol.* **A2**, 957–960.
37. Treglia, G., Desjonqueres, M. C., Sparjaard, D. *et al.* (1981) Study of the W(TA) core-level shift induced by the adsorbtion of oxygen. *J. Phys.* **C14**, 3463–3473.
38. Wertheim, G. K. and Crecelius, G. (1978) Divalent surface state of metalic samarium. *Phys. Rev. Lett.* **40**, 813–816.
39. Rosengren, A. and Johansson, B. (1980) Calculated transition-metal surface core-level binding energy shifts. *Phys. Rev.* **B22**, 3706–3709.
40. Tomanek, D., Kumar, V., Holloway, S., and Benneman, K. H. (1982) Semiempirical theory for surface core-level shifts. *Solid State Commun.* **41**, 273–279.
41. Heimann, P., Van der Veen, J. F., and Eastman, D. E. (1981) Structure-dependent surface core-level shifts for the Au(111), (100), (110) surfaces. *Solid State Commun.* **38**, 595–598.
42. Johanson, B. and Märtensson, N. (1980) Core-level binding energy shifts for the metalic elements. *Phys. Rev.* **21**, 4427–4457.
43. Brocksch, H. -J., Tomanek, D., and Benneman, K. H. (1983) Calculation of surface core-level shifts in intermediate-valence compounds. *Phys. Rev.* **B27**, 7313–7317.
44. Castle, J. E. (1977) Use of X-ray photoelectron in corrosion science. *Surf. Sci.* **68**, 583–602.
45. Akimov, A. G. (1984) Initial oxidation state of metallic systems. *Poverkhnost* **3**, 5–22 (in Russian).
46. Janik-Czahor, M. (1979) AES investigations of passive films on iron and iron base alloys. *Corrosion* **35**, 360–366.
47. Hashimoto, K., Asami, K., and Teramoto, K. (1979) An X-ray photoelectron spectroscopic study on the role of molybdenum in increasing the corrosion resistance of ferric stainless steels in HCl. *Corros. Sci.* **19**, 3–14.
48. Hashimoto, K. and Asami, K. (1979) An X-ray photoelectron spectroscopic study of the passivity of ferritic 19 Cr stainless steels in 1N HCl. *Corros. Sci.* **19**, 251–266.
49. Hashimoto, K. and Asami, K. (1979) XPS-study of surface film on nickel alloys in hot concentrated NaOH. *Corros. Sci.* **19**, 427–435.
50. Hashimoto, K., Naka, M., Asami, K., and Masumoto, T. (1979) An X-ray photoelectron spectroscopy study of the passivity of amorphous Fe–Mo alloys. *Corros. Sci.* **19**, 165–170.
51. Rozenfeld, I. L., Kazansky, L. P., Akimov, A. G., and Frolova, L. V. (1979) X-ray photoelectron study of inorganic inhibitors on the surface of iron. *Zashch. Met.* 349–352 (in Russian).
52. Burstein, G. T. and Newman, R. C. (1980) Surface films on corroded binary copper alloys. *Appl. Surf. Sci.* **4**, 162–173.
53. Okamoto, Y., Carter, W. J., and Hercules, D. M. (1979) A study of the interaction of Pb–Sn solder with O_2, H_2O and N_2O by X-ray photoelectron and Auger spectroscopy. *Appl. Spectrosc.* **33**, 287–295.
54. Castle, J. E. (1983) Simultaneous electron and X-ray analysis and its application in corrosion science. *J. Vac. Sci. Technol.* **1**, 1013–1020.
55. Akimov, A. G., Gagarin, A. P., Dorofeyev, V. G. *et al.* (1982) Study of the oxide film composition forming on the surface of chrome-nickel steel under the influence of pulse heating in air. *Poverkhnost* **1**, 141–143 (in Russian).
56. McInture, N. S., Zetaruk, D. G., and Owen, D. (1978) XPS study of initial growth of oxide films on inconel-660 alloy. *Appl. Surf. Sci.* **2**, 55–73.

57. Olefjord, J. (1975) Application of ESCA to oxide films formed on stainless steels at intermediate and high temperatures. *Matl. Sci.* **9**, 263–268.
58. Akimov, A. G., Rozenfeld, I. L., Kazansky, L. P. and Machavariani, G. V. (1978) X-ray photoelectron study of iron–chrome and iron–chrome/nickel alloy oxidation. *Izv. Akad. Nauk. SSSR Ser. Khim.* 1482–1486.
59. Leygraf, C., Hultquist, G., Ekelund, S., and Erikson, J. C. (1974) Surface composition studies of the (100) and (110) faces of monocrystalline $Fe_{0.84}Cr_{0.16}$. *Surf. Sci.* **46**, 157–176.
60. Kadar, I., Köver, I., and Cserny, I. (1980) XPS measurement of oxide layers on stainless steel surfaces. *ATOMKI Közl.* **22**, 29–30.
61. Akimov, A. G. and Machavariani, G. V. (1980) X-ray photoelectron spectroscopy study of oxidation of $Fe_{0.75}Cr_{0.25}$ alloy. *Zashchi. Met.* 32–34 (in Russian).
62. Storp, S. and Holm, R. (1977) ESCA investigation of oxide layers on some Cr-containing alloys. *Surf. Sci.* **68**, 10–19.
63. Olefjord, J. (1975) ESCA studies composition profile of low temperature oxide formed on chromium steels 1. Oxidation in dry oxygen. *Corros. Sci.* **15**, 687–696.
64. Akimov, A. G., Zimin, P. A., and Rozenfeld, I. L. (1982) Study of the oxidation of aluminium bronze. *Poverkhnost* **1**, 119–121 (in Russian).
65. Akimov, A. G. and Dagurov, V. G. (1982) Study of surface composition during the oxidation of titanium and its alloys. *Poverkhnost* **1**, 75–80 (in Russian).
66. Leygraf, G. and Hultquist, G. (1976) Initial oxidation stages on Fe–Cr (100) and Fe–Cr(110) surfaces. *Surf. Sci.* **61**, 69–84.
67. Müssig, H. J., and Adolphi, B. (1980) On the initial interaction of oxygen with the (100) stainless steel surface studied by AES and LEED. *ATOMKI Közl.* **22**, 27–28.
68. Betz, G., Vehner, G. K., and Toth, L. (1974) Composition VS depth profiles obtained with Auger electron spectroscopy of air-oxidized stainless steel surfaces. *J. Appl. Phys.* **45**, 5312–5319.
69. Winograd, N., Baitinger, W. E., Amy, J.W., and Munarin, J. A. (1974) X-ray photoelectron spectroscopic studies of interaction in multicomponent metal and metal oxide thin films. *Science*, **184**, 565–567.
70. Holm, R. and Storp, S. (1976) Surface analysis of Ni/Al alloys by X-ray photoelectron spectroscopy. *J. Electron Spectrosc. Rel. Phen.* **8**, 139–147.
71. Lee, W. Y. and Eldridge, J. (1977) Oxidation studies of permalloy films by quartz crystal microbalance, AES and XPS. *Trans. Electrochem. Soc.* **124**, 1747–1751.
72. Nefedov, V. I., Pozdeev, P. P., Dorfman, V. F., and Pypkin, B. N. (1980) X-ray photoelectron study of thin evaporated films of Fe/Ni alloy. *Surf. Interface Anal.* **2**, 26–30.
73. Brundle, C. R., Silverman, E., and Madix, R. J. (1979) Oxygen interaction with Ni/Fe surface (1) LEED and XPS studies of Ni 76% (Fe 24%) (100). *J. Vac. Sci. Technol.* **16**, 474–477.
74. Nefedov, V. I., Lunin, V. V., and Chulkov, N. G. (1980) Dependence of surface composition of Zr–Ni intermetallic compounds. *Surf. Interface Anal.* **2**, 207–211.
75. Fidler, R., Chulkov, N. G., Solovetsky, Y. I., Lunin, V. V., and Nefedov, V. I. (1983) X-ray photoelectron study of catalyst surfaces on the basis of $ZrCo\ H_{2.88}$ and $HfCo\ H_{2.88}$ hydrides. *Poverkhnost* **2**, 49–51 (in Russian).
76. Fidler, R., Chulkov, N. G., Kuznetsova, N. N., Lunin, V. V., and Nefedov, V. I. (1983) X-ray photoelectron study of catalysts on the basis of $ZrNiH_{2.8}$ and $HfNiH_{2.8}$ hydrides. *Poverkhnost* **2**, 111–115 (in Russian).
77. Fidler, R., Nefedov, V. I., Chulkov, N. G. *et al.* (1984) X-ray photoelectron study of the surface oxidation of the HfCo intermetallide. *Poverkhnost* **3**, 87–91 (in Russian).

78. Nefedov, V. I. (1984) X-ray photoelectron study of the surface of inorganic materials. *Neorg. Mat.* **20**, 942–948 (in Russian).
79. Castele, J. E. and Nasserian-Riabi, M. (1975) Oxidation of Cu–Ni alloys. 1. XPS study of interdiffusion. *Corros. Sci.* **15**, 537–543.
80. Van Ooij, W. E. (1977) Role of XPS in study and understanding of rubber to metal bonding. *Surf. Sci.* **68**, 1–9.
81. Maroie, S., Haemers, G., and Verbist, J. J. (1984) Surface oxidation of polycrystalline α- and β-brass as studied by XPS and AES. *Appl. Surf. Sci.* **17**, 463–476.
82. Hegde, M. S., Sumpath, T. S., and Mallya, R. M. (1983) A study of surface segregation and oxidation of Cu–Mn alloys by XPS and AES. *Appl. Surf. Sci.* **17**, 97–106.
83. Bernabai, U., Covallini, M., Bombara, G. *et al.* (1980) The effects of heat treatment and implantation of aluminium on the oxidation resistance of Fe–Cr–Al–Y–alloys. *Corros. Sci.* **20**, 19–25.
84. Shabanova, I. N., Yermakov, A. N., Trapeznikov, V. A., and Shur, Y. S. (1974) Dependence of saturation magnetization of nickel aerosol powders on the state of surface particles studied by photoelectron spectroscopy. *Fiz. Metal. Metalloved.* **38**, 314–322 (in Russian).
85. Veremeyenko, M. D., Alenchikova, I. F., and Nefedov, V. I. (1984) X-ray photoelectron study of the fluoridized surface of Fe–Ni alloys. *Poverkhnost* **3**, 55–63 (in Russian).
86. Nefedov, V. I. Veremeyenko, M. D., and Alenchikova, I. F. (1984) X-ray photoelectron study of iron fluorides. *Zh. Neorg. Khim.* **29**, 2422–2424 (in Russian).
87. Ranke, W., Xing, Y. R., and Shen, G. D. (1982) Orientation dependence of oxygen adsorption on a cylindrical GaAs sample. *Surf. Sci.* **122**, 256–274.
88. Schwartz, G. P., Gualter, G. J., Kammlott, G. W., and Schwartz, B. (1979) An X-ray photoelectron spectroscopy study of native oxides on GaAs. *J. Electrochem. Soc.* **126**, 1737–1749.
89. Grunthaner, P. J., Vasquez, R. P., and Grunthaner, F. J. (1980) Chemical depth profiles of the GaAs/native oxide interface. *J. Vac. Sci. Technol.* **17**, 1045–1051.
90. Brundle, C. R. and Seybold, D. (1979) Oxygen interaction with GaAs surface. An XPS/UPS study. *J. Vac. Sci Technol.* **16**, 1186–1190.
91. Lindau, I., Pianetta, P., Spicer, W. E. *et al.* (1978). Oxygen adsorption and the surface electronic structure of GaAs(110). *J. Electron Spectrosc. Rel. Phen.* **13**, 155 –160.
92. Jacobi, K. and Ranke, W. (1976) Oxidation and annealing of GaP and GaAs(111)-faces studied by AES and UPS. *J. Electron Spectrosc. Rel. Phen.* **8**, 225–238.
93. Chye, P. W., Su, C. Y., Lindau, I. *et al.* (1979) Oxidation of ordered and disordered GaAs(110). *J. Vac Sci. Technol.* **16**, 119–1194.
94. Su, C. Y., Lindau, I., Skeath, P. R. *et al.* (1980) Oxygen adsorption on the GaAs(110) surface. *J. Vac. Sci. Technol.* **17**, 936–941.
95. Chye, P. W., Su, C. Y., Lindau, I., *et al.* (1979) Photoemission studies of the initial stages of oxidation of GaSb and InP. *Surf. Sci.* **88**, 439–460.
96. Pianetta, P., Lindau, I., Garner, C. M., and Spicer, W. E. (1978) Chemisorption and oxidation studies of (110) surface of GaAs, GaSb and InP. *Phys. Rev.* **B18**, 2792–2806.
97. Ebina, A., Asano, K., Suda, Y., and Takahashi, T. (1980) Oxidation properties of II–VI compound surfaces studied by low-energy electron-loss spectroscopy and 21 eV photoemission spectroscopy. *J. Vac. Sci. Technol.* 1074–1079.
98. Chen, M., Batra, I. P., and Brundle, C. R. (1979) Theoretical and experimental investigations of the electron structure of oxygen on silicon. *J. Vac. Sci. Technol.* **16**, 1216–1220.

99. Grunthaner, F. J., Grunthaner, P. J., Vasquez, R. R. *et al.* (1979) High resolution X-ray photoelectron spectroscopy as a probe of local atomic structure. *Phys. Rev. Lett.* **43**, 1683–1686.
100. Ishizaka, A., Iwata, S., and Kamigaki, Y. (1979) Si–SiO_2 interface characterization by ESCA. *Surf. Sci.* **84**, 355–374.
101. Barr, T. L. (1983) An XPS study of Si as it occurs in adsorbents, catalysts, and thin films. *Appl. Surf. Sci.* **15**, 1–15.
102. Hübner, K. (1977) Chemical bond and related properties of SiO_2 core level shifts in SiO_x. *Phys. Status Solidi* **A42**, 501–509.
103. Bechstedt, F. (1979) Chemical shift of Si2P core level in SiO_x. Calculation of relaxation contribution. *Phys. Status Solidi* **B91**, 167–176.
104. Mende, G., Finster, J., Flamm, D., and Schulze, D. (1983) Oxidation of etched solicon at room temperature. *Surf. Sci.* **128**, 169–175.
105. Hollinger, G. and Himpsel, F. J. (1983) Oxygen chemisorption and oxide formation on Si(111) and Si(100) surfaces. *J. Vac. Sci. Technol.* **A1**, 640–645.
106. Hecht, M. H., Grunthaner, F. J., Pianetta, P. *et al.* (1984) Electron escape depth variation in thin SiO_2 films. *J. Vac. Sci. Technol.* **A2**, 584–587.
107. Skeath, P., Lindau, I., Pianetta, P. *et al.* (1979) Photoemission study of the interaction of Al with a GaAs(110) surface. *J. Electron Spectrosc. Rel. Phen.* **17**, 259–265.
108. Coad, J. P. (1972) X-ray photoelectron and Auger spectroscopic analysis of steel surfaces. Preprint, United Kingdom Atomic Energy Authority, Harwell AERE-R J-944.
109. Shabanova, I. N., Kutyin, A. B., Smirnov, L. V., and Trapeznikov, V. A. (1976) Use of X-ray photoelectron spectroscopy for study of antimony segregation during the brittling of steel. *Fiz. Met. Metalloved.* **42**, 318–322 (in Russian).
110. Ustinovshchikov, Y. I., Shabanova, I. N., Sapukhin, V. A., and Trapeznikov, V. A. (1977) Brittling of alloyed steels during tempering. *Fiz. Met. Metalloved.* **44**, 336–334 (in Russian).
111. Shabanova, I. N. and Trapeznikov, V. A. (1977) Use of X-ray photoelectron spectroscopy in studying and analysing surface fractures of brittled materials. *Fiz. Elektron Tverd. Tela.*, Izhevsk, Udmurt University, 191–200 (in Russian).
112. Trapeznikov, V. A. (1982) An XPS study of the surface layers of metals and alloys. *Poverkhnost* **1**, 18–26 (in Russian).
113. Eggert, B. and Panzer, G. (1982) Electron spectroscopic study of phosphorus segregated to α-iron surface. *Surf. Sci.* **118**, 345–368.
114. Panzer, G. and Eggert, B. (1984) The bonding state of sulfur segregated to α-iron surfaces and on iron sulfide surfaces studied by XPS, AES and ELS. *Surf. Sci.* **144**, 651–664.
115. Ho, P., Mitchell, D. F., and Graham, M. J. (1983) Surface and grain boundary segregation related to the temper embrittlement of a $2\frac{1}{4}$ Cr–1 Mo steel. *Appl. Surf. Sci.* **15**, 108–119.
116. Rats, Y. V., Pushin, V. P., Romanova, R. R. *et al.* (1978) Study of ageing aluminium-silver alloy by X-ray photoelectron spectroscopy and electron diffraction spectroscopy. *Fiz. Met. Metalloved.* **45**, 795–802 (in Russian).
117. Chappell, R. and Stoddart, C. (1974) Auger electron spectroscopy study of float glass surfaces. *Phys. Chem. Glasses* **15**, 130–136.
118. Baitinger, W., French, P., and Swarts, E. (1980) Characterization of tin in the bottom surface of float glass by ellipsometry and XPS. *J. Non-crystall. Solids* **38/39**, 749–754.
119. Colombin, L., Charlies, H., Jelli, A., Debres, G., and Verbist, J. (1980) Penetration of tin in the bottom surface of float glass: a synthesis. *J. Non-crystall. Solids* **38/39**, 551–556.

120. Budd, S. (1975) ESCA examination of tin oxide coatings on glass surfaces. *J. Non-crystall. Solids* **19**, 55–64.
121. Kleshchevnikov, A. M., Milovanov, A. P., Moiseyev, V. V., Nefedov V. I., Permyakova, T. V., Portnyagin V. I., and Sheshukova, G. Y. (1982) X-ray photoelectron study of the surface of sodium silicate glasses. *Fiz. Khim. Stekla* **8**, 742–744 (in Russian).
122. Kleshchevnikov, A. M. (1983) In-depth element concentration profiles in solids on the basis X-ray photoelectron spectroscopy data. Synopsis of thesis for the degree of candidate of sciences. Moscow (in Russian)
123. Kleshchevnikov, A. M., Milovanov, A. P., Moiseyev, V. V., Nefedov, V. I., Portnyagin, V. I., and Shashkina, G. A. (1983) X-ray photoelectron spectroscopic study of sodium silicate glass after ion exchange processing. *Fiz. Khim. Stekla* **9**, 622–628 (in Russian).
124. Hanert, M. and Rauschenbach, B. (1983) A study of the surface layers of silicate glass. *Fiz. Khim. Stekla* **9**, 696–703 (in Russian).
125. Rauschenbach, B. (1983) Segregation on Silicatglasoberfläche. Silikattechnik **34**, 299–301.
126. Smets, B. and Lommen, T. (1981) SIMS and XPS investigation of the leaching of glasses. *Verres Refract.* **35**, 84–90.
127. Smets, B. and Lommen, T. (1982) The leaching of sodium-aluminosilicate glasses studied by secondary ion mass spectrometry. *Phys. Chem. Glasses* **23**, 83–89.
128. Hench, L. (1977) Physical chemistry of glass surfaces. Eleventh International Congress on Glass, Part 1, Prague, pp. 343–369.
129. Müller, W., Hähnert, M., Berg, V., and Brümmer, O. (1978) Einbau und analytischer Nachweis von Erdalkalien in Glasoberfächen bei Ionenaustauschpvoressen. *Silikat Technik* **29**, 205–207.
130. Puglisi, O., Torrisi, A., and Marletta, G. (1984) XPS investigation of the effects induced by the silanization on real glass surfaces. *J. Non-crystall. Solids* **68**, 219–230.
131. Brückner, R., Chun, H., and Goretzki, H. (1978) ESCA on alkali silicate and soda aluminosilicate glasses. *Glastech. Ber.* **51**, 1–7.
132. Brückner, R., Chun, H., Goretzki, H., and Sammet, M. (1980) XPS measurements and structural aspects of silicate and phosphate glasses. *J. Non-crystall. Solids* **42**, 49–60.
133. Smets, B. and Krol, D. (1984) Group III ions in sodium silicate glass. Part 1. X-ray photoelectron spectroscopy study. *Phys. Chem. Glasses* **25**, 113–118.
134. Smets, B. and Gommen, T. (1981) The structure of germanosilicate glasses studied by XPS. *J. Non-crystall. Solids* **46**, 21–32.
135. Bach, H. (1983) Oberflächen-und Dünnschichtanalysen an Glasoberfächen und Oberflächenbelägen. *Glastech. Ber.* **56**, 63–70.
136. Milovanov, A. P., Moiseyev, V. V., and Portnyagin, V. I. (1985) Modern methods of surface analysis when studying glass. *Fiz. Khim. Stekla.* **11**, 3–29 (in Russian).
137. Nefedov, V. I., Firsov, M. N., Vasiljev, V. Ju., and Sotnikova, E. V. (1985) ESCA-investigation of amorphous Ni–Mo alloys. *Poverkhnost* **4**, 117–122.
138. Nefedov, V. I., Pozdejev, P. P. *et al.* (1986) ESCA and X-Ray spectroscopic investigations of amorphous Ni–Mo alloys. *J. Electron Spectrosc. Rel. Phen.* **40**, 11–26.
139. Nefedov, V. I., Salyn, Ya. V., Solozhenkin, P. M., and Pulatov, G. Yu. (1980) X-ray photoelectron study of surface compounds formed during floation of minerals. *Surf. Interface Anal.* **2**, 170–172.
140. Mielczarski, J., Werfel, F., and Suoninen, E. (1984) XPS-studies of xanthate with copper surfaces. *Appl. Surf. Sci.* **17**, 160–174.
141. Nefedov, V. I., Salyn, Ya. V., Makeyev, V. A., and Zelenov, V. V. (1982) Surface

composition of Ag–Au alloys and native gold. *Poverkhnost* **1**, 144–149 (in Russian).
142. Nefedov, V. I., Zhavoronkov, N. M., Machavariani *et al.* (1982) The variation of silver content in native gold partides as revealed by ESCA. *Phys. Chem. Minerals* **8**, 193–196.
143. Nefedov, V. I., Aseyev, P. I., Kleshchevnikov, A. M. *et al.* (1982) Study of diffusion and segregation in Au/Ag thin layers. *Poverkhnost* **1**, 71–73 (in Russian).
144. Nefedov, V. I., Salyn, Ya. V., Makeyev, V. A., and Zelenov, V. V. (1981) Surface composition of native gold and Au/Ag alloys. *J. Electron Spectrosc. Rel. Phen.* **24**, 11–17.
145. Zhavoronkov, N. M., Nefedov, V. I., Machavariani, G. V. *et al.* (1982) Dependence of silver content in the surface layer of native gold on the deposit's genetic class and type. *Dokl. Akad. Nauk SSSR* **263**, 1459–1462 (in Russian).
146. Vinogradov, A. P., Nefedov, V. I., Urusov, V. S., and Zhavoronkov, N. M. (1971) X-ray photoelectron study of lunar regolith from the Seas of Fertility and Tranquility. *Dokl. Akad. Nauk SSSR* **201**, 957–960 (in Russian).
147. Vinogradov, A. P., Nefedov, V. I., Urusov, V. S., and Zhavoronkov, N. M. (1972) ESCA-investigation of lunar regolith from seas of Fertility and Tranquility. Proceedings of the Third Lunar Science Conference Vol. 2, pp. 1421–1427.
148. Housley, R. M. and Grant, R. W. (1975) ESCA studies of lunar surface chemistry. Proceedings of the Sixth Lunar Science Conference Vol. 3, pp. 3269–3275.
149. Housley, R. M. and Grant, R. W. (1976) ESCA studies of the surface chemistry of Lunar fines. Proceedings of the Seventh Lunar Science Conference Vol. 1, pp. 881–889.
150. Dikov, Y. P., Nemoshkalenko, V. V., Alyoshin, V. G. *et al.* (1977) Reduced titanium in Lunar regolith. *Dokl. Akad. Nauk SSSR* **234**, 176–179 (in Russian).
151. Dikov, Y. P., Bogatikov, O. A., Alyoshin, V. G. *et al.* (1977) Reduced silicon in lunar regolith. *Dokl. Akad. Nauk SSSR* **235**, 1410–1420 (in Russian).
152. Nefedov, V. I., Sergushin, N. P., Salyn, Y. V. *et al.* (1977) X-ray photoelectron studies of iron and the surface characteristics of lunar regolith from the Sea of Crises. *Geokhimia* 1516–1523 (in Russian).
153. Vinogradov, A. P., Nefedov, V. I., Urusov, V. S., and Zhavoronkov, N. M. (1972) X-ray photoelectron study of metallic iron in lunar regolith. *Dokl. Akad. Nauk SSSR* **207**, 433–436 (in Russian).
154. Housley, R. M. and Grant, R. W. (1977) An XPS (ESCA) study of lunar surface alteration profiles. Proceedings of the Eighth Lunar Science Conference **3**, 3885–3899.
155. Dikov, Yu. P., Bogatikov, O. A., Barsukov, V. L. *et al.* (1978) Some features of the main element conditions in surface layers of the regolith particles of the Luna automatic stations samples: X-ray photoelectronic spectroscopy studies. Proceedings of the Ninth Lunar Planet Science Conference Vol. 3, pp. 2111–2124.
156. Nefedov, V. I., Urusov, V. S., and Zhavoronkov, N. M. (1972) Differences in the concentration of main elements on the surface and in the bulk lunar regolith particles. *Dokl. Akad. Nauk SSSR* **207**, 698–701 (in Russian).
157. Khirny, Y. M. and Solodovnikov, A. P. (1974) Increased corrosion-resistance of metals irradiated by helium ions. *Dokl. Akad. Nauk SSSR* **214**, 82–83 (in Russian).
158. Nefedov, V. I., Sergushin, N. P., Salyn, Ya. V. *et al.* (1980) X-ray photoelectron study of iron and surface characteristics of lunar regolith from the Sea of Crises. In: Lunar regolith from the Sea of Crises. Moscow, Nauka, p. 318–324 (in Russian).
159. Hirvonen, J. K. (1978) Ion implantation in thribology and corrosion science. *J. Vac. Sci. Technol.* **15**, 1662–1668.
160. Holland, H. D., Blackburn, T. R., Ceasar, G. P., and Quirk, R. F. (1979) X-ray

photoelectron spectrometric and gas exchange evidence for surface oxidation of Martian regolith analogues by ultraviolet radiation. *Life Sci. Space Res.* **17**, 65–76.
161. Baun, W. L. (1980) Applications of surface analysis techniques to study of adhesion. *Appl. Surf. Sci.* **4**, 291–306.
162. Richter, K., Kley, G., Robbe, J. *et al.* (1978) Röntgenphotoelektro nenspektroskopische Untersuchung der Metall-Polymer-Phasen Grenze in Verbundmaterialien. *Z. Chem.* **18**, 390.
163. Richter, K., Gühde, J., Loeschcke, I. *et al.* (1978) Röntgenphoto elektronenspektroskopische Untersuchung an Bruchoberflachen von Kupfer-Polymer-Verbundfolien. *Plast. Kauc.* **25**, 202–206.
164. Wyatt, D. M., Gray, R. C., Carver, J. C., and Hercules, D. M. (1974) Studies of polymeric bond failure on aluminium surface by X-ray photoelectron spectroscopy. *Appl. Spectrosc.* **28**, 439–446.
165. Burkstrand, J. M. (1982) Chemical interaction at polymer–metal interfaces and the correlation with adhesion. *J. Vac. Sci. Technol.* **20**, 440–441.
166. Burkstrand, J. M. (1979) Copper–polyvinŷl alcohol interface: a study with XPS. *J. Vac. Sci. Technol.* **16**, 363–365.
167. Marletta, G., Pignataro, S., and Sancisi, G. (1984) ESCA investigation of a failure case in brass coated steel cords used in radial tyres. *Appl. Surf. Sci.* 390–400.
168. Cooperthwait, R. C. (1980) The study of radiation-induced chemical damage at solid surface using photoelectron spectroscopy. *Surf. Interface Anal.* **2**, 17–25.
169. Cooperthwait, R. G. (1980) X-ray photoelectron spectroscopic evidence for noval surface nitrogen species in irradiated $NaNO_3$. *J. Chem. Soc. Commun.* **7**, 320–321.
170. Kabešova, M., Sramko, T., Gazo, J. *et al.* (1978) Thermal properties of thiocyanatocopper (II) complexes with picolines and lutidines. *J. Therm. Anal.* **13**, 55–64.
171. Sasaki, T., Williams, R. S., Wong, J. S., and Shirley, D. A. (1978) Radiation damage studies by X-ray photoelectron spectroscopy. I. Electronirradiated $LiNO_3$ and Li_2SO_4. *J. Chem. Phys.* **68**, 2718–2724.
172. Sasaki, T., Willaims, R. S., Wong, J. S., and Shirley, D. A. (1978) Radiation damage studies by X-ray photoelectron spectroscopy. II. Electron irradiated Li_2CrO_4 and Li_2WO_4. *J. Chem. Phys.* **69**, 4374–4380.
173. Sasaki, T., Williams, R. S., Wong, J. S., and Shirley D. A. (1979) Radiation damage studies by X-ray photoelectron spectroscopy. III. Electron-irradiated halates and perhalates. *J. Chem. Phys.* **71**, 4601–4610.
174. Brinen, J. S. and Wilson, L. A. (1972) Elucidation of damage mechanism of synthetic sodalites by ESCA. *J. Chem. Phys.* **56**, 6256–6257.
175. Wagner, C. D. (1984) The contribution of Bremsstrahlung to radiation damage by ESCA. *Surf. Interface Anal.* **6**, 90–91.
176. Fasana, A. and Braicovich L. (1982) Chemically driven diffusion on Pd in the surface region of Zn(0001). *Surf. Sci.* **120**, 239–250.

Supplement

This supplement contains a compilation of experimental data of various authors on core level binding energies in elements from Fuggle and Maertensson's paper[a]. The types of spectrometer (ABCDE) cited in Table I–IV are described below.

A—ESCA-3 spectrometer (Vacuum Generators) at Strathclyde University; B—high vacuum spectrometer designed by Vacuum Generators for the Technical University in Munich; C and D—Hewlett-Packard spectrometers in Upsalla and at the Max Planck Institute in Stuttgart; E—high vacuum spectrometer designed by the firm Kratos for the Institut fur Festkörperforschung in Julich.

Table I
Binding energies of elements 4, 11–14[a]

Element	Spectrometer	1s	2s	2p
4Be	A	111.8	—	—
11Na	E	1070.8	63.5	30.4
12Mg	A	1303.0	88.6	49.6
13Al	A		117.9	72.8
14Si	A			99.4
	E		150.3	99.1

[a]Fuggle, J. C. and Maertensson, N. (1980) Core level binding energies in metals. *J. Electron Spectrosc. Rel. Phen.* **21**, 275–281.

Table II
Binding energies of elements 20–31

Element	Spectrometer	2s	$2p_{1/2}$	$2p_{3/2}$	3s	$3p_{1/2}$	$3p_{3/2}$	3d
20Ca	A	438.8		346.4	44.4	25.4		
	D	438.0	349.7	346.0	44.2			
22Ti	A	560.7	460.3	453.6	58.7	32.7		
	C	561.0	460.1	453.9	58.6	32.5		
23V	A	626.6	519.7	512.0	66.3	37.1		
	C	626.8	519.8	512.1	66.2	37.2		
24Cr	A	695.7	584.0	573.9	74.0	42.0		
	C	696.3	583.6	574.3	74.3	42.3		
25Mn	A	768.7	649.8	638.6	82.3	47.2		
	C	769.4	649.9	638.8	82.4	47.2		
26Fe	A		719.8	706.8	91.2	52.8		
	C	844.6	719.9	706.7	91.3	52.6		
27Co	A		792.9	777.8	100.9	58.8		
	D	925.1	793.4	778.3	101.0	59.0		
28Ni	A	1008.4	869.8	852.6	110.9	68.0	66.2	
	C	1008.9	870.1	852.8	110.6		66.2	
29Cu	A	1096.7	952.3	932.5	122.5	77.3	75.1	
31Ga	A		1143.2	1116.4	159.5	103.5	100.0	18.7

Table III
Binding energies of elements 38–56

Element	Spectro-meter	3s	$3p_{1/2}$	$3p_{3/2}$	$3d_{3/2}$	$3d_{5/2}$	4s	$4p_{1/2}$	$4p_{1/2}$	$4p_{3/2}$	$4d_{3/2}$	$4d_{5/2}$
38Sr	D	358.7	280.3	270.0	136.0	134.2	38.9	21.6		20.3		
39Y	A				157.7	158.8						
	C					158.8						
40Zr	A		343.5	329.9	181.0	178.8						
	C	430.3	343.4	329.7	181.2	178.8	50.6	28.5		27.1		
41Nb	C	466.6	376.1	360.6	205.0	202.3	56.4	32.6		30.8		
42MO	C	506.3	411.6	394.0	231.1	227.9	63.2	37.6		35.5		
44Ru	B	586.2	483.3	461.2	284.1	279.9	75.0	46.5		43.2		
	C	586.1	483.7	461.5	284.2	280.0	75					
45Rh	C	628.1	521.3	496.5	311.9	307.2	81.4	50.5		47.3		
46Pd	A	672.0	559.9	532.3	340.5	335.3	87.6	55.7[a]		50.9[a]		
	C	671.2	559.9	532.3	340.4	335.1	87.6		52[a]			
47Ag	A	718.9	603.7	573.0	373.9	367.9						
	B	718.7	603.6	572.7	373.9	367.9	96.7	63.7[a]		58.3[a]		
	C	719.4	604.1	573.3	374.2	368.2	97.4		60[a]			
48Cd	D	772.0	652.6	618.4	411.9	405.2	109.8		63.9[a]		11.7	10.7
49In	A	827.1	703.0	665.1	451.4	443.9	122.9				17.7	17.0
	C	827.2	703.3	665.4	451.4	443.9	122.9		73.5[a]		17.7	16.8
50Sn	A	884.0	756.4	714.5	493.1	484.8	136.9				24.8	23.8
	C	885.3	756.6	714.7	493.3	484.9	137.2		83.6[a]		25.0	24.0
51Sb	C	946	812.7	766.4	537.5	528.2	153.2		95.6[a]		33.3	32.1
52Te	C	1006	870.8	820.0	583.4	573.0	169.4		103.3[a]		41.9	40.4
56Ba	D				795.5	780.2	253.5			178.6	92.6	89.9

[a] One-electron approximation for the level is incorrect.

Table IV
Binding energies of elements 56–92

Element	Spectro-meter	4s	$4p_{1/2}$	$4p_{3/2}$	$4d_{3/2}$	$4d_{5/2}$	$4f_{5/2}$	$4f_{7/2}$	5s	$5p_{1/2}$	$5p_{3/2}$	$5d_{3/2}$	$5d_{5/2}$	6s	$6p_{1/2}$	$6p_{3/2}$
56Ba	D	253.5		178.6	92.6	89.9			30.3	17.0	14.8					
72Hf	C		438.2	380.7	220.0	211.5	15.9	14.2	64.2		29.9					
73Ta	C	563.4	463.4	400.9	237.9	226.4	23.5	21.6	69.7		32.7					
74W	B	593.8	490.0	423.6	255.8	243.4	33.6	31.4	76.6							
	C	494.3	490.8	423.7	256.0	243.5	33.5	31.3	75.5		36.8					
75Re	C	625.4	518.7	446.8	273.9	260.5	42.9	40.5	83.0	45.6	34.6					
76Os	C	658.2	549.1	470.7	293.1	278.5	53.4	50.7			44.5					
77Ir	C	691.1	577.8	495.8	311.9	296.3	63.8	60.8			48.0					
78Pt	A	725.0	609.0	519.2	331.6	314.6	74.5	71.2			51.7					
	C	725.7	609.2	519.5	331.5	314.6	74.4	71.1			51.7					
79Au	A	761.9	642.5	546.2	353.0	335.0	87.5	83.7			57.2					
	C	762.3	642.9	546.4	353.3	335.2	87.7	84.0	107.2	74.2	57.2					
80Hg	D	802.2	680.2	576.6	378.2	358.8	104.0	99.9	127	83.1	64.5	9.6	7.8			
81Tl	A	846.0	720.9	609.2	405.6	384.9	122.2	117.8		94.5	73.5					
	C	846.4	720.0	609.7	405.8	385.0	122.2	117.7		94.6	73.4	14.7	12.5			
82Pb	A	891.7	761.9	643.4	434.5	412.3	141.7	136.9		106.2	83.2	20.8	18.2			
	C	891.8	761.9	643.5	434.1	412.0	141.6	136.8		106.6	83.3	20.6	18.0			
83Bi	A		805.0	678.8	463.9	440.0	163.3	157.0		118.8	92.4	26.9	23.8			
	C	939	805.4	678.8	464.0	440.1	162.2	156.9		119.1	92.8	26.9	23.8			
90Th	A			966.4	712.1	675.2	342.4	333.1		—[a]	—[a]	92.5	85.4	41.4	24.5	16.6
92U	A			1043.0	778.3	736.2	388.2	377.4		—[a]	—[a]	102.8	94.2	43.9	26.8	16.8

[a] One electron approximation for the level is incorrect.